# Women Who Did Science

# Women Who Did Science

Five remarkable women whose discoveries changed our understanding of the world

R. D. Bennett

Published by Snowhill America

www.snowhillamerica.com



First Edition: 2026

Paperback ISBN: 979-8-9926797-9-3

Hardcover ISBN: 979-8-9968072-1-5

eBook ISBN: 979-8-9968072-0-8

This book is intended for educational and informational purposes only. The author and publisher have made every effort to ensure the accuracy of the information presented, but they do not make any representations or warranties regarding the completeness, accuracy, or suitability of the contents. Historical interpretations and scientific understanding may evolve as new information becomes available.

All portrait illustrations are original commissioned artwork based on historical reference materials and are not reproductions of any single photograph.

Cover by germancreative

Illustrations by Osmo

Internal formatting by Ebooks Experts

# WOMEN WHO DID SCIENCE

**Five remarkable women whose discoveries changed our understanding of the world.**

**Meet the Women**

**Marie Curie**

**Jane Goodall**

**Ada Lovelace**

**Rosalind Franklin**

**Vera Rubin**

Curiosity has changed the world. Throughout history, extraordinary women have challenged accepted ideas, overcome barriers, and expanded the boundaries of human knowledge.

Their stories remind us that one determined person can change the world, and that curiosity, courage, and perseverance can inspire generations to follow.

# Table of Contents

# INTRODUCTION

When people think of great scientific discoveries, they often think of famous names such as Isaac Newton, Albert Einstein, Charles Darwin, or Stephen Hawking. Yet throughout history, many women have also transformed our understanding of the world, often while facing obstacles that their male colleagues never encountered.

For centuries, universities, scientific societies, laboratories, and observatories excluded women. Many institutions denied them formal education, prevented them from attending lectures, or discouraged them from pursuing careers in science altogether. Some women worked without pay, while others saw their discoveries credited to colleagues, supervisors, or institutions. Despite these barriers, women continued to ask questions, conduct experiments, challenge assumptions, and expand human knowledge.

This book tells the stories of five remarkable women whose work changed science forever.

Some, like Marie Curie and Jane Goodall, became internationally famous during their lifetimes. Others, such as Vera Rubin, received widespread recognition only years after their most important discoveries. Ada Lovelace and Rosalind Franklin achieved full recognition only after their deaths. Each followed a different path, but they shared a common trait: an unwavering curiosity about how the world works.

In these pages, you will meet:

- **Marie Curie (1867–1934)**, whose pioneering work on radioactivity transformed physics and medicine.
- **Jane Goodall (1934–2025)**, whose observations of chimpanzees revolutionized our understanding of animal behavior and humanity's place in nature.
- **Ada Lovelace (1815–1852)**, who envisioned that computers could one day perform tasks far beyond simple calculation.

- **Rosalind Franklin (1920–1958)**, whose meticulous research helped reveal the structure of DNA, the molecule that carries the genetic instructions for life.
- **Vera Rubin (1928–2016)**, whose observations provided some of the strongest evidence for the existence of dark matter, one of the greatest mysteries in modern astronomy.

These women did far more than make discoveries. They challenged accepted ideas, opened doors for future generations, and showed that scientific progress depends on determination, creativity, and persistence.

This book is not only about their lives. It is also about the science that they pursued. Understanding what these women discovered, and why those discoveries mattered, allows us to appreciate the true scale of their contributions. Their work continues to influence medicine, technology, genetics, conservation, astronomy, and countless other fields today.

Many depict science as a series of sudden breakthroughs or moments of inspiration. In reality, it is usually a long process of observation, experimentation, failure, and persistence. The stories in this book reveal that scientific discovery is rarely straightforward. It requires patience, courage, and a willingness to keep asking questions when the answers remain uncertain.

The women in this book did exactly that. They explored the unknown, challenged what others thought was possible, and expanded our understanding of the world around us. Their discoveries changed the course of science, and their stories continue to inspire new generations of scientists, thinkers, and explorers.

Perhaps the greatest lesson they leave us is that curiosity has no boundaries. Gender, background, or circumstance does not limit it. It begins with a question, grows through perseverance, and has the power to change the world.

# MARIE CURIE

(1867-1934)

## Physicist and Chemist

First woman to win a Nobel Prize and the only person to win Nobel Prizes in two different sciences.

Original portrait illustration by Osmo, based on a historical reference photograph (c. 1898).

"Be less curious about people and more curious about ideas."
—*Marie Curie*

Marie Curie entered adulthood in a society that denied women access to higher education, but she refused to accept those boundaries. Her relentless pursuit of knowledge carried her across borders and into the heart of scientific discovery. Her work would transform our understanding of matter and energy.

## Childhood

Marie was born Maria Salomea Skłodowska in Warsaw, Poland, on November 7, 1867. Maria was the youngest of five children. Her father, Władysław Skłodowski, was a professor of mathematics and physics, and her mother, Bronisława Boguska, was the principal of a girls' school.

In 1867, Poland was under the control of the Russian Empire after the failed January Uprising of 1863-64, and Russian authorities in Russian-controlled Poland restricted Polish-language education, limited access to higher learning, and monitored schools for political dissent. However, Maria's parents believed strongly in educating girls, and they created a home culture that treated their daughters' intellect as equal to that of their sons. At home, Maria and her sisters, Bronislawa (Bronia) and Helena (Hela), had access to books, experiments, and their father's scientific instruments, just like her brothers, Zofia (Zosia) and Józef. This was unusual, as people did not prioritize girls' education at this time. Therefore, while girls could attend primary school, secondary schools for girls were rare. The secondary school Maria attended taught in Polish, despite the Russian language requirement. During inspection visits, they had to pretend, and Maria, the best student, often stood up and recited Russian lessons. This act convinced inspectors that the school followed the law.

During this time, life was hard. In 1876, typhus took Maria's older sister, Zofia (Zosia). In 1878, Maria's mother died from tuberculosis when she was ten, and her father lost his job because of his pro-Polish political views. When Maria's father realized he could not support his family, they took in lodgers to survive.

Yet Maria never stopped studying. She graduated from secondary school at 15 years old with a gold medal as first in her class from a Russian-run school. However, Maria then spent a year recovering from depression and exhaustion. In the summer of 1883, Maria stayed for two months with a former student of her mother. She also spent nearly a year in the countryside in the Lublin region, where one of her uncles lived, and received a warm welcome.

## The Flying University

Between 1885 and 1886, Marie joined the Flying University (Uniwersytet Latający), a secret network of classes for women. Russian rule prevented women and politically restricted groups from accessing formal higher education, so in the late 19th and early 20th centuries, in Poland, an underground educational network known as the Flying University (later the more formal Society for Scientific Courses) emerged. The classes "flew" from apartment to apartment to avoid police detection, and leading Polish scholars committed to intellectual freedom taught lessons. The Flying University offered rigorous instruction in science, mathematics, literature, and philosophy and was a vital pathway for women like Maria and Bronia seeking advanced study.

Maria and her older sister, Bronia, dreamed of leaving Poland and getting a degree, as women could not attend the University of Warsaw, but they lacked the financial resources to pay for more schooling. Maria didn't give up and made a deal with her sister: she'd support Bronia through med school, and Bronia would pay her back once she graduated.

So, Maria worked as a governess in Warsaw and nearby towns to help support the family and Bronia's medical studies in Paris, teaching children in wealthy homes. Maria worked as a tutor and governess for nearly five years. She spent her free time studying, reading about physics, chemistry, and math, and working part-time to cover her expenses. During this time, money was often so tight that Maria would faint from hunger, but she never stopped learning.

## Heartbreak and Determination

At 19, Maria worked for the prominent Żorawski family in Szczuki, Poland. She taught the family's young son, developing a friendship with his older sister, and organizing literacy programs for the peasant children on the estate. There, she met Paulin Kazimierz Żorawski, the family's son and a brilliant math student at the University of Warsaw. Maria and Paulin fell in love, but Paulin's parents disapproved of the match, citing Maria's family's financial struggles. Ultimately, Paulin's parents denied their wish to marry, citing the bad financial conditions of the Curie family, and they pressured Paulin to end the relationship. This heartbreak devastated Maria, and people frequently point to it as the cause of her leaving Warsaw for Paris.

Back in Warsaw, Maria learned chemistry and physics in a laboratory at the Museum of Industry and Agriculture, managed by her cousin Józef Boguski. He was a major player in Warsaw's science underground and had worked with Dmitri Mendeleev, the Russian who created the periodic table by sorting elements by their atomic weight and chemical traits. Boguski, who had a deep understanding of analytical chemistry and laboratory techniques, recognized Maria's talent and gave her hands-on training in chemical experiments.

Another mentor, Napoleon Milicer, taught her analytical chemistry. Milicer supported young students who wanted to pursue scientific study despite political and educational barriers. Through these mentors, Maria gained the skills and confidence that would eventually lead to her world-changing discoveries.

During this time, Maria continued to take part in Flying University lectures between 1889 and 1890. However, as women could not enroll in universities, including the University of Warsaw, under the Russian occupation, Maria followed in the footsteps of her sister Bronia.

## Arrival in Paris

In November 1891, at 24 and at the urging of her sister, Maria followed Bronia to Paris, leaving her home and family behind. With almost no money, Maria carried little more than a suitcase, her notebooks, and a fierce determination to learn. Paris was the place where she would turn her passion for discovery, setting the stage for her legendary career.

Maria stayed briefly with her sister Bronia, who lived with her fiancé (later husband), Kazimierz Dłuski, also of Polish origin, having met him at medical school in Paris. They helped Maria settle in and enrolled at the Sorbonne, one of Europe's most prestigious universities.

The horse bus took almost an hour to get Maria to the university from the Dluskis' place, so she got her own apartment instead of staying with her sister permanently. Maria resented the lost time, not to mention the money wasted on carfare, and wanted her independence and a quiet place to study.

In addition, staying with the Dluskis during her early weeks in Paris meant taking part regularly in the active Polish exile community in the city. Maria's father warned her that this could hurt her career opportunities back home and potentially endanger her relatives.

Bronia and Kazimierz helped Maria find a room, an unheated attic room in the Latin Quarter of Paris. In the 1890s, universities, laboratories, libraries, bookshops, and inexpensive student rooms filled the Latin Quarter. Its cafés and narrow streets buzzed with international students, writers, and scientists, creating an atmosphere of constant debate and discovery. For a young scholar with limited means, the Latin Quarter offered both opportunity and community, an ideal setting for Maria's early scientific training. Thus, after a few weeks of living with her sister, Maria moved to the Latin Quarter in Paris.

Bronia and Kazimierz initially helped cover Maria's early living expenses, including rent. It was freezing in winter, stifling in summer, and her room was barely big enough for a bed and a desk. Maria lived frugally during this time, often surviving on bread, butter, and tea, and sometimes fainting from hunger. However, these harsh conditions built her discipline and resilience.

---

**HISTORICAL SPOTLIGHT**

### Bronisława and Kazimierz Dłuski

Bronisława Skłodowska Dłuska (1865–1939), Maria Curie's older sister, Bronia, became one of the first Polish women physicians after completing her medical studies in Paris. A specialist in gynecology and obstetrics, she later co-founded and directed the Maria Skłodowska-Curie Institute of Oncology in Warsaw, one of Europe's earliest cancer centers. Her leadership in women's health and medical education made her a respected figure in Polish medicine.

Her husband, Kazimierz Dłuski (1855–1930), was a physician and a prominent social activist. He played a major role in public health and political life and was the founding president of the Tatra Volunteer Rescue Service (TOPR), an organization that remains central to mountain rescue efforts in Poland.

---

## Studying at the Sorbonne

The Sorbonne, the historic center of the University of Paris, was one of Europe's most respected academic institutions when Maria arrived in 1891. Founded in the 13th century, it had a long tradition of rigorous scholarship, and its Faculty of Science attracted leading physicists, chemists, and mathematicians of the time. For Maria, the Sorbonne offered world-class instruction and access to advanced scientific facilities, providing the foundation for her early research and future discoveries.

When Maria registered at the Sorbonne in Paris, she signed her name as "Marie" to seem more French. Marie immersed herself in physics, mathematics, and chemistry. She spent her days in classes and her nights studying. With no social life and no distractions, within a year, she rose to the top of her physics class. In 1893, Marie was first in her class and earned her first degree, the Licence ès Sciences Physiques, equivalent to a modern bachelor's degree in physics.

Marie received the Alexandrovitch Scholarship for Polish students studying abroad in recognition of her talent. The scholarship enabled Marie to remain in Paris and finance her remaining classes, allowing her to continue her education.

In 1893, she worked with Gabriel Lippmann, a leading French physicist and future Nobel laureate. Lippmann advanced optics and electrical instrumentation, creating tools that supported early research like Marie's studies of steel magnetization, a problem important for industry and electricity. Lippmann is best known for inventing interferential color photography, the first method to produce true-color images without pigments or dyes, a breakthrough that earned him the 1908 Nobel Prize in Physics.

In 1894, Marie obtained her Licence ès Sciences Mathématiques, equivalent to a modern bachelor's degree in mathematics. She learned from the best mathematicians of her day, Paul Appell and Paul Painlevé. Appell, a famous French mathematician, made important contributions to classical mechanics, differential equations, and the theory of functions. As dean of the Sorbonne's Faculty of Science, he played a major role in modernizing scientific instruction and expanding laboratory facilities, earning widespread respect as both a researcher and a teacher.

Paul Painlevé gained recognition for his innovative contributions to differential equations and singularities. The Painlevé equations remain central to modern mathematical physics because they describe universal nonlinear behavior across quantum theory, statistical mechanics, integrable systems, and many physical applications. Painlevé also entered public life, serving twice as Prime Minister of France. Together, they created the rigorous, imaginative, and intellectually charged environment in which Marie's scientific career began.

## Post Graduate Research

Marie's goal after graduating was to take a teacher's diploma and return home to Poland to support Polish education, especially for girls. When Marie tried to return to work in Poland in 1894, Krakow University denied her a place because she was a woman. As a result, Marie returned to Paris to pursue her PhD.

In 1894, Lippmann was so impressed by Marie that he gave her full authority over a project to study how different types of steel behave when magnetized, especially how their magnetic properties change with composition and heat treatment.

This was an industrially important problem in the 1890s because steel was becoming essential for railways, machinery, and electrical devices. In addition, manufacturers needed to understand how to make steel with predictable magnetic behavior, and France wanted to improve its industrial competitiveness. Since Marie knew little about this topic, she asked Józef Wierusz-Kowalski, a Polish scientist who was an expert in it, for help.

Marie's first major scientific project in Paris focused on the magnetization of tempered steel, a topic at the center of late-19th-century physics. Working at a borrowed bench in Lippmann's laboratory at the Sorbonne, she measured how steel's magnetic strength changed with composition, treatment, and temperature. This research required the type of high-precision instruments developed by physicists like Lippmann. Her work helped clarify how magnetic order behaves in solid materials and introduced her to the rigorous methods of French experimental physics.

Wierusz-Kowalski recognized Marie needed better laboratory access and equipment, as this work required extremely precise instruments, and so introduced Marie to Pierre Curie, a 35-year-old professor at École Municipale de Physique et de Chimie Industrielles de la Ville de Paris (ESPCI), the School of Physics, in 1894.

**SCIENCE SPOTLIGHT**

### Magnetization

Magnetization describes how a material becomes magnetic when its internal atomic moments begin to align. Each atom behaves like a tiny bar magnet, but in most materials, these moments point in random directions and cancel out. Applying an external magnetic field causes regions called magnetic domains to rotate, aligning more of their moments. As additional domains align, the material's overall magnetic strength increases. Once all domains point the same way, the material reaches saturation, its maximum magnetization. Some materials lose this alignment when the field is removed, while others, like permanent magnets, keep it. Understanding magnetization helps explain ferromagnetism, the Curie temperature, and how magnetic materials store information.

## Pierre Curie

Pierre Curie, who was born in 1859 and raised in Paris, attended school at home until he was 16. Thanks to homeschooling, Pierre could spend more time exploring his interests in math and physics, and his father, a physician, encouraged him. When he was 16, Pierre started studying at the University of Paris's Faculty of Sciences. He finished his physics degree by 18, a testament to his intelligence and hard work. After gaining his degree, Pierre worked as a laboratory assistant at the Sorbonne. In 1882, at 23 years old, he became a supervisor at ESPCI.

Before Pierre met Marie in 1894, he had already completed pioneering work on magnetism. He had also investigated the phenomenon of materials losing their magnetism at specific temperatures, a concept known as the Curie Point, which now forms a fundamental part of physics. He also developed some of the most sensitive electrometers and balances available, such as the Curie balance (used to measure minute magnetic forces), high-precision electrometers, and piezoelectric measuring instruments. Piezoelectricity occurs in certain crystals, including quartz, tourmaline, and Rochelle salt. Applying pressure, like squeezing, stretching, or twisting, generates an electric charge in these crystals. The greater the pressure, the more the crystal distorts, and the more electricity the crystal generates. Such crystals are now used in microphones, electronic apparatus, and clocks.

In 1888, Pierre filed a patent for a precision aperiodic (non-oscillating) balance manufactured by the Société Centrale des Produits Chimiques (SCPC), Central Society of Chemical Products. However, it was not this instrument that Marie needed for her steel magnetization project. What she required were the highly sensitive electrometers Pierre had developed with his brother Jacques, which made him the ideal person to assist her with research. Wierusz-Kowalski recognized that Pierre and Marie had compatible research interests and thought they would get along scientifically.

---

**SCIENCE SPOTLIGHT**

## Pierre Curie-Magnetism and Piezoelectricity

Pierre Curie identified a precise mathematical relationship showing that every magnetic substance loses its magnetism above a characteristic temperature, now called the Curie point, and formulated Curie's Law, a cornerstone of magnetism. His research helped define the major categories of magnetic behavior: ferromagnetism (strong, permanent magnetism), paramagnetism (weak, temporary magnetism), and diamagnetism (repulsion by magnetic fields).

Pierre also co-discovered piezoelectricity (electricity generated by pressure) in 1880 with his brother Jacques, revealing that certain crystals generate an electric charge when compressed. In 1881, they succeeded in demonstrating the reverse phenomenon: the tiny contraction or expansion of a crystal subjected to electric voltages. Beginning in 1884, Pierre developed a series of reflections and writings on symmetry in crystallography and physics, extending its application beyond the description of shapes. The six articles he published established what he called "the principle of symmetry." These breakthroughs made him one of Europe's leading experimental physicists even before his partnership with Marie.

---

**SCIENCE SPOTLIGHT**

### The Curie Point

The Curie point is the temperature at which a ferromagnetic material loses its permanent magnetism. Below this temperature, atoms within materials such as iron align into stable magnetic domains, giving the material strong and lasting magnetic properties. As temperature increases, thermal energy disrupts this internal order. When the Curie point is reached, the magnetic domains lose their alignment, and the material becomes paramagnetic, responding only weakly to external magnetic fields.

Pierre's research showed that every ferromagnetic material has its own characteristic Curie temperature, a value determined by its atomic structure. His work laid the foundation for modern studies of magnetism and remains important in physics, materials science, and engineering today.

## Magnetization

Marie's work on the magnetization of tempered steel involved measuring:

- **Magnetic susceptibility:** How strongly a steel sample becomes magnetized in an applied magnetic field.
- **Magnetic hysteresis:** How steel "remembers" magnetization, a property crucial for early electrical devices.
- **The effect of impurities and alloy composition:** Different steels (carbon steel, nickel steel, etc.) showed distinct magnetic behaviors.
- **The effect of temperature:** Heating steel changes its internal structure and, therefore, its magnetism.

Working with Pierre, Marie measured the magnetic properties of a wide range of substances. She applied the relationship Pierre had discovered, later formalized as Curie's Law, showing that the magnetic susceptibility of a paramagnetic material varies inversely with temperature. In simple terms, Curie's Law means that a paramagnetic material becomes less magnetic as it gets hotter. While Pierre is credited with formulating Curie's Law in 1895, Marie significantly contributed to its experimental verification and extension, confirming that many materials followed the same temperature-dependent behavior.

Curie's Law became a cornerstone of statistical mechanics and magnetism, explaining why paramagnets lose magnetization when heated. Marie's careful study of magnetization helped establish the generality of Curie's Law across different classes of materials, including those with complex atomic structures. Her work deepened the understanding of paramagnetism and showed that the temperature dependence Pierre identified applied far beyond the substances originally tested.

## Pierre and Marie Curie

As Wierusz-Kowalski predicted, Marie and Pierre got along well, sharing a love of science and a zest for adventure. Her senior by about a decade, Pierre, aged 35, was still living quietly with his parents. He was shy, introverted, and wholly absorbed in his scientific work. Until this time, he had devoted his life entirely to science, his family, and his quiet routines at ESPCI. Marie was the first person to awaken in him a sense of companionship beyond the laboratory. In Marie, he found an equal with a comparable devotion to science. They both enjoyed cycling and quickly formed a bond.

Marie encouraged Pierre to write up his long-running research on magnetism and temperature as a doctoral thesis, *Magnetic Properties of Bodies at Various Temperatures*. The thesis summarized fifteen years of work on different forms of magnetism and presented his discovery that magnetic susceptibility varies inversely with temperature, now known as Curie's Law. He also showed that ferromagnetic materials lose their magnetism above a critical transition point, the Curie temperature.

Although Pierre had already made major scientific contributions in several fields, he had never completed a doctorate; in France, the process was especially time-consuming. Persuaded by Marie and his father, he finally agreed to submit the thesis. He received his PhD in March 1895, along with a promotion to professor at the ESPCI.

Marie would eventually find a small laboratory space at the Société d'Encouragement pour l'Industrie Nationale, an industrial research society, while her relationship with Pierre grew from mutual respect to love.

When Pierre proposed to Marie in 1895, she did not accept right away. She was worried about being far from her country and family. She returned to Poland for a vacation later that year with the thought of moving back to Poland, but Pierre wrote to her many times while she was away and convinced her to return. Three months later, they married on July 26, 1895, in Sceaux, France, a year after they had first met. Marie's father and sisters, who were present, met Pierre's family. Ever practical, Marie insisted that her wedding dress be a dark color instead of the traditional white so she could wear it in their laboratory as well.

The couple used the money they had received as a wedding present to purchase two identical men's bicycles. These were state-of-the-art models at the time, equipped with pneumatic tires. Pierre's handlebar had a travel bag, while Marie's had a flower garland. Pierre wore a sports jacket and pants; Marie, a straw hat, a puffed-shoulder jacket, and knee-length culottes for the many long rides they took together. Every evening, they would stop randomly at an unknown hotel.

After the marriage in 1895, Marie gained French nationality. The couple initially lived in modest rooms in Paris before moving in 1898 to a small house at 108 Boulevard Kellermann, near the Parc Montsouris in Paris. Here they continued their cycling excursions, often taking day trips on their bicycles. They spent summer holidays in Auvergne, Brittany, and Normandy and frequently visited Pierre's parents in Sceaux, about fifteen kilometers from Paris, sometimes cycling there when the weather was good.

## Discovering Radioactivity

On December 28, 1895, Wilhelm Conrad Roentgen announced the discovery of a new type of radiation, highly penetrating rays which he named

X-rays. The image of his wife's hand skeleton, placed between the emitting tube and a photographic plate, had an immediate worldwide impact, with doctors in all Western countries started using the X-ray apparatus described by Wilhelm to perform scans of their own as soon as mid-January 1896. After Henri Poincaré lectured at the Académie des Sciences on January 20, 1896, he suggested that Henri Becquerel investigate phosphorescence. Certain materials exhibit phosphorescence, meaning they absorb energy and then gradually emit it as visible light, which causes them to glow long after the original light source is gone.

In March 1896, at the Natural History Museum in Paris, Becquerel exposed many samples, including uranium salts, to sunlight to study whether phosphorescent materials might also emit X-rays. Becquerel placed phosphorescent materials on top of photographic paper wrapped in thick black paper and placed them in bright sunlight for several hours. He showed that the developed plate showed shadows of the objects. On February 27, 1896, unable to expose a phosphorescent uranium salt to the sun because it was a cloudy day, he placed the uranium on top of a fresh sheet of photographic paper at the back of a drawer. Some days later, he opened the drawer and developed the photographic film. Becquerel found that the uranium salt had spontaneously emitted a penetrating ray capable of leaving an impression on the photographic plate. As no external energy was required for these rays, this made them different from X-rays. Becquerel found out fast that different uranium salts, glowing or not, all did the same thing. This proved that the radiation was in uranium. He named the rays from uranium salts 'uranic rays' and showed their ability to electrify the air. Later analysis revealed that these rays were a mixture of beta and gamma radiation, capable of ionizing air and passing through thin materials. This unexpected discovery marked the first evidence of radioactivity.

In 1896, Marie passed her teacher's diploma, coming first in her group. In 1897, she concluded several investigations into the magnetic properties of steel on behalf of an industrial association. During this time, Marie used part of her award from the Society for the Encouragement of National Industry to repay the Polish scholarship that had supported her studies four years earlier. Though she was under no obligation to return the funds, she did so to ensure another deserving Polish student could receive the same opportunity. Deciding after some time to continue doing research, Marie looked around for a subject for a doctoral thesis.

Marie picked uranic rays as her PhD subject in late 1897. After only a few days in Lippmann's lab, she found that thorium gave off the same rays as uranium. Her consistent research on various chemical compounds, like salts, surprisingly showed that the radiation strength stayed the same, regardless of the compound. It depended only on the amount of uranium or thorium present. Chemical compounds of the same element generally have very different chemical and physical properties: for example, one uranium compound is a dark powder, while another is a transparent yellow crystal, but what was different was the amount of radiation they gave off, as it related only to the amount of uranium the chemical compounds contained. Unlike X-rays, these uranium rays seemed to come directly from the uranium itself. Marie quickly realized something revolutionary: chemical or physical reactions did not cause radioactivity but was a property of the atom itself.

This was a completely new idea. Marie then went through the whole periodic system available at the time and found that only uranium and thorium gave off this radiation.

---

**SCIENCE SPOTLIGHT**

## X-Rays

Discovered by Wilhelm Röntgen in 1895, X-rays are a form of high-energy electromagnetic radiation that has higher energy and wavelengths much shorter than those of visible light. To create X-rays, external energy is required. They are neutral particles, meaning magnetic fields cannot bend them, and this characteristic distinguishes them from uranium rays based on charge. Typically, an X-ray tube produces them by accelerating electrons through a potential difference. It then directs them to a target material, such as tungsten.

X-rays can also expose photographic plates and ionize air, revealing a new kind of invisible radiation that immediately transformed medical imaging.

The penetrating power of X-rays allows them to pass through soft tissues while being absorbed by denser materials like bone or metal. As a result, these parts appear as white areas in the image. Softer body parts like the lungs and heart are easier for X-rays to pass through, showing up as darker areas.

---

**SCIENCE SPOTLIGHT**

## Atoms, Molecules, Electrons, and their Chemical Bonds

Atoms are the basic building blocks of matter, combining to form the molecules that make up everything around us. Each atom has a dense central nucleus made of protons and neutrons, surrounded by electrons that occupy regions of space called electron shells. These electrons are constantly in motion, and their arrangement determines how an atom behaves chemically. When two or more atoms share or exchange electrons, they form molecules, the stable units that make up everything from water and air to proteins and plastics.

Chemical bonds describe the ways atoms connect to form stable molecules. Electrons shared, transferred, or pooled between atoms create these connections. In covalent bonds, atoms share pairs of electrons, creating strong, directional links found in water, oxygen, and most organic molecules. Ionic bonds happen when an atom hands over electrons to another, making charged particles that then attract each other, like with table salt. Metallic bonds occur when many atoms share a loose "sea" of electrons, giving metals their conductivity and flexibility. The type of bond depends on each atom's tendency to gain or lose electrons, a property called electronegativity.

The nucleus is the compact center of an atom, containing nearly all its mass. Protons, which carry a positive charge, and neutrons, which carry no charge, make up the nucleus. The strong nuclear force holds these particles together, a powerful attraction that overcomes the natural repulsion between positively charged protons. The number of protons defines the element itself; carbon always has six, oxygen eight, while the number of neutrons can vary, creating different isotopes of the same element. Although tiny compared to the whole atom, the nucleus determines an atom's stability, radioactivity, and how it transforms during nuclear reactions. Changes in the nucleus release far more energy than chemical reactions, which involve only electrons.

## The Curie Method of Radioactivity

Marie first measured radioactivity using only a quadrant electrometer and an ionization chamber. Marie placed the sample inside the chamber, where ionizing radiation knocked electrons off gas molecules, creating ion pairs. An electric field drew these charges to the electrodes, producing a tiny current proportional to the radiation's intensity. If a substance were radioactive, it would make the air between the electrometer plates conductive, allowing the current to be measured.

However, Marie's laboratory notes show that measurements made with just the electrometer and chamber were not sufficiently reliable. For precise quantitative results, she used Pierre's piezoelectric quartz electrometer, developed by Pierre and his brother Jacques in 1880. Based on the piezoelectric effect, it allowed Marie and Pierre to measure extremely weak electrical currents with exceptional precision. They had discovered that quartz emits an electric charge when compressed, and that the greater the pressure, the greater the charge.

During a measurement, the charge from the sample moved the electrometer's light spot along a scale. Marie then gently raised a weight attached to the quartz blade, generating a known compensating charge. A chronometer measured how long the quartz charge balanced the sample's charge. The shorter the time, the stronger the radioactivity. By comparing the known quartz charge with the unknown charge from the sample, the Curies could determine radioactivity with remarkable accuracy.

Marie performed these delicate measurements in the semidarkness of an unheated room to avoid temperature fluctuations. She described the apparatus as: "The two ends of the blade [of quartz] are secured between a set of jaws.... the lower jaw is fitted with a plate on which weights can be placed.... This traction has the effect of releasing quantities of electricity on both sides of the blade.... which are then collected."

Her colleague Catherine Chamié recalled Marie's joy when repeated measurements agreed within extremely small limits. This technique, later known as the Curie Method of Radioactivity, remained in use in the Curie laboratory into the 1950s.

Using this method, Marie examined many elements and compounds and showed that only thorium, like uranium, emitted spontaneous radiation. She established that radioactivity was an atomic property and coined the term radioactivity. Marie also showed that certain minerals, including pitchblende from Bohemia, emitted far more radiation than their uranium content could explain.

The objective of the Curie Method is to measure the number of electric charges produced, which is proportional to a sample's radioactive emissions. This insight laid the foundation for measuring radioactivity, isolating radioactive isotopes, and understanding radioactive decay and half-life, opening the door to the discovery of new elements and the birth of atomic physics.

## SCIENCE SPOTLIGHT

### How Scientists Measure Radiation Today

Modern radiation detection relies on tools that translate invisible particles and rays into measurable signals. Geiger counters are the most familiar: they detect individual ionizing events as charged particles pass through a gas-filled tube, producing the characteristic clicks that show radiation intensity. Scientists use scintillation detectors for more precise measurements. These detectors contain crystals that emit tiny flashes of light when radiation strikes them. These flashes are then amplified and counted electronically.

To identify exactly which radioactive isotopes are present, laboratories turn to gamma spectroscopy, a technique that measures the unique energy "fingerprints" of gamma rays emitted by different nuclei. For individuals who work with radiation, like medical professionals, researchers, or nuclear power technicians, dosimeters offer a continuous record of their exposure over time. This ensures their safety, permitting the responsible use of radiation in medical fields, scientific research, and industrial applications.

---

**SCIENCE SPOTLIGHT**

## Radioactivity

Radioactivity is the natural process by which an unstable atomic nucleus transforms itself into something more stable. Instead of remaining fixed, certain nuclei release energy in the form of alpha particles, beta particles, or gamma rays. Each type of emission carries away a bit of mass or energy, gradually reshaping the nucleus. As this happens, the atom literally changes into a different element, step by step, through a chain of transformations.

This idea was astonishing in the late nineteenth century. Before Marie Curie's work, scientists considered atoms indivisible and eternal, as the unchanging building blocks of matter. Radioactivity overturned that belief. It showed that some atoms are inherently unstable and that their transformations follow precise, measurable laws. The discovery revealed a hidden dynamism inside matter itself and opened the door to modern nuclear science, from understanding the age of the Earth to developing medical imaging and cancer treatments.

---

**SCIENCE SPOTLIGHT**

## Understanding Half-Life

The half-life of a radioactive substance is the time it takes for half of its atoms to undergo decay and transform into something else. Instead of disappearing all at once, radioactive materials fade in a steady, predictable rhythm. After one half-life, half the original atoms remain; after two half-lives, only a quarter remain; after three, an eighth, and so on. This pattern continues no matter how much material you start with.

Radioactive decay follows an exponential pattern because each atom behaves independently. At any moment, every atom has the same probability of decaying, whether it has existed for a day or a century. This constant probability produces a smooth, mathematical curve rather than a straight line, one reason half-life became such a powerful tool in early nuclear science.

Different elements decay at dramatically different rates. Polonium-210, the isotope Marie Curie first identified, has a half-life of only 138 days. Its activity drops quickly, which is why Marie noticed her early polonium samples weakening over time. Radium-226, by contrast, has a half-life of about 1,600 years. Its radioactivity remains essentially unchanged over a human lifetime, making it far easier to study and isolate, one reason radium became the centerpiece of the Curies' research.

Today, the concept of half-life underpins many modern technologies. Carbon-14 dating uses the predictable decay of carbon isotopes to determine the age of ancient artifacts and fossils. In medicine, specialists use short-lived isotopes for imaging and cancer therapy, which deliver targeted radiation that quickly fades from the body. Whether in archaeology, geology, physics, or healthcare, half-life remains one of the most elegant and practical ideas to emerge from the early study of radioactivity.

## Searching for New Elements

Marie's systematic studies of many chemical compounds produced a surprising result: the strength of the radiation depended only on the amount of uranium or thorium present, not on the compound itself. Uranium compounds can differ greatly in appearance and chemical properties, yet the radiation they emitted was determined solely by their uranium content. From this, Marie concluded that a substance's chemical composition and molecular structure did not influence its ability to radiate.

She reviewed the entire periodic table as it was known and discovered that only uranium and thorium emitted spontaneous radiation. Around the same time, the German physicist Gerhard Carl Schmidt independently reported that thorium also possessed this property.

Marie next turned her attention to natural ores containing uranium and thorium. She obtained samples from geological museums, including specimens from Professor Alfred Lacroix's extensive mineral collection at the Museum of Natural History in Paris, where Becquerel had conducted some of his investigations. Among these samples was pitchblende, a uranium oxide mineral and the most important ore of uranium.

When Marie tested it, she discovered that pitchblende exhibited four to five times more radioactivity than its uranium content alone could explain. She hypothesized that the ore must contain an unknown element far more radioactive than uranium. Intrigued by her findings, Pierre set aside his own research on crystals and symmetry to join her investigation. Guided by Marie's conviction that radioactivity was an atomic property, the pair began a search that would soon lead to the discoveries of polonium and radium. During this period, Marie also introduced the term radioactivity to describe this remarkable phenomenon.

---

**SCIENCE SPOTLIGHT**

**Ore**

An ore is a naturally occurring rock or mineral that contains enough valuable metal or element to be mined and processed profitably. What makes a rock an ore is not just the presence of a metal, but the concentration, accessibility, and economic value of that metal at a given time. Ores often form through geological processes that concentrate elements, such as heat, pressure, hydrothermal fluids, or weathering, into deposits richer than ordinary crustal rock. Common examples include iron ore, copper ore, and uranium ore such as uraninite. Once extracted, ores are crushed and chemically treated to separate the desired metal from the surrounding material. Because market prices and technology change, a rock that is not considered an ore today may become one in the future if its metal becomes more valuable or easier to extract.

---

## The Mystery of Pitchblende

The Curies required tons of the costly ore pitchblende to prove their theory about polonium and radium, as Marie understood that the residues, following uranium's removal, held the new elements. Pitchblende is a black, radioactive, uranium-rich mineral, also known as uraninite, and is the primary ore of uranium.

However, people knew then that the Joachimsthal mine in Bohemia had left large slag heaps of pitchblende in the surrounding forests. The Joachimsthal mine was an old silver and uranium mine, which, as a byproduct of the silver mining, produced pitchblende as waste.

A sample of pitchblende from the Joachimsthal mine arrived from Bohemia to the Curies, and they discovered that the slag was even more active than the original mineral. The Austrian Academy of Sciences later made several tons of pitchblende available to them.

Marie did the chemical separations, and Pierre took the measurements after each step. For Marie, it was physically demanding work, having to move 20 kg of pitchblende per batch. Before she could separate the pitchblende, she had to remove any pine needles or other materials from it; then she had to undertake the work of separation. "Sometimes I had to spend a whole day stirring a boiling mass with a heavy iron rod nearly as big as myself. I would be broken with fatigue at day's end," Marie wrote.

Working tirelessly with Pierre, they:

- Dissolved pitchblende in acid
- Separated its components using chemical techniques
- Measured radiation with precise instruments, including Pierre's sensitive electrometers

Through painstaking work, they confirmed both Polonium, a new element named after Marie's homeland, Poland, emitting intense radiation, and Radium, even more radioactive, later isolated in a pure form after years of labor.

---

**HISTORICAL SPOTLIGHT**

### Joachimsthal Uranium Mine

The Joachimsthal (Jáchymov) uranium mine in the Ore Mountains became one of the most historically significant mining sites in Europe. Once a silver mining hub in the 16th century, the town's prosperity declined until prospectors discovered a resistant black mineral, pitchblende, which complicated silver purification. This "nuisance" ore later proved extraordinary. In 1789, Martin Heinrich Klaproth identified uranium in Joachimsthal pitchblende, which sparked a new industry for uranium-based pigments in glass and ceramics. By the late 19th century, demand for these pigments collapsed, leaving vast piles of discarded ore. In 1898, when Marie and Pierre Curie extracted polonium and radium from Joachimsthal pitchblende, it triggered a global rush for radium as a medical wonder. Because the ore contained only trace amounts of radium, Joachimsthal became the world's primary source, fueling a brief but intense boom until richer uranium deposits elsewhere ended its dominance.

---

## Polonium

Marie and Pierre found that the strongest activity in pitchblende came from pieces that contained bismuth or barium. Continuing her analysis of the bismuth fractions, Marie discovered that each time she removed some bismuth, the remaining residue grew even more active. By June 1898, they had isolated material roughly 300 times more radioactive than uranium.

On July 18, 1898, Marie and Pierre published a brief communication to the Académie des Sciences titled *"Sur une nouvelle substance fortement radio-active contenue dans la pechblende"* (On a new strongly radioactive substance contained in pitchblende). This paper was the first time they used the term "radioactivité" (radioactivity) in print and was the first time that the Curies publicly declared that radioactivity was not merely a property of uranium but a sign of new elements.

In the communication, they reported that the substance which they extracted from pitchblende behaved chemically like bismuth, yet was far more radioactive than uranium. They suggested that this could indicate the presence of a new metal. If confirmed, they wrote, they proposed to name it polonium, after the homeland of "one of us," a bold gesture at a time when Poland did not exist as an independent nation.

Although the Curies already suspected the presence of a second new element in the barium fractions, they left Paris soon after for a long-planned holiday in Auvergne with little Irène.

After announcing polonium in July 1898, Marie continued refining the bismuth residues and confirmed that the new substance's radioactivity diminished over time, an early insight into radioactive decay. While she and Pierre were never able to isolate pure polonium, her continued work soon revealed that an even more intensely radioactive substance was present in the barium fraction of pitchblende. As her work continued, she realized that an even more intensely radioactive substance was present in the barium fraction of pitchblende.

## Radium

On December 26, 1898, Marie, Pierre, and Gustave Bémont announced to the Académie des Sciences that they had strong evidence for a second new radioactive element in pitchblende, one that behaved chemically like barium but emitted extraordinarily intense radiation. With spectroscopic confirmation from Eugène Anatole Demarçay, they proposed the name radium. Radium emissions were hundreds of times stronger than uranium, and samples produced a characteristic blue radioluminescent glow in their surroundings, not because the element itself emitted light, but because its intense radiation excited the surrounding air and glass.

It took Marie over three years to isolate even a tenth of a gram of radium chloride. After thousands of crystallizations, Marie finally, from several tons of the original material, isolated one decigram of almost pure radium chloride and determined radium's atomic weight as 225. Polonium was harder, as its half-life was only 138 days, making it almost impossible to isolate.

The first paper that was presented at a meeting on July 18, 1898, announced they'd found a new radioactive element, and the Curies called it polonium, after Marie's home country. At the December 26 meeting, the Curies presented a second paper that announced radium's discovery.

Processing tons of pitchblende from 1898 to 1902, the Curies' incredible work earned them more than just valuable centigrams of radium. They published thirty-two papers together or alone during that time. In one of these papers, the Curies showed that radium exposure destroyed diseased, tumor-forming cells faster than healthy cells.

In early June 1903, the Royal Institution in London invited Marie and Pierre as guests. Pierre then lectured major scientists in England, including Lord Kelvin, a leading British physicist, who sat next to Marie during the lecture as a sign of respect. Lord Kelvin established the absolute temperature scale that now bears his name (kelvin) and made foundational contributions to thermodynamics, including the formulation of the second law and the concept of absolute zero. His stature in European science made his eventual acceptance of the Curies' results an important moment in the recognition of their work.

While Marie was present at the lecture, she could not speak because of the Royal Institution's policies at that time, which excluded women from lecturing. Pierre showed the crowd how radium could quickly affect photographic plates that were wrapped in paper, how it emitted heat, and how it glowed when it was semi-dark. While lecturing, Pierre brought up the idea of using radium for cancer treatment. He showed this by wrapping radium salts on his arm for ten hours, which resulted in a burn-like injury that left a lasting gray scar after 52 days. Pierre's hands were shaking and scarred as he dealt with the radium, and he ended up spilling some. Fifty years after this lecture, scientists discovered radioactivity on the premises, and certain surfaces had to be cleaned.

So, after that successful lecture, Marie wrote up her findings in her doctoral thesis on June 25, 1903. Later, two of the three people on the examination committee won Nobel Prizes; her old teacher, Lippmann, received one for physics in 1908, and Ferdinand Frédéric Henri Moissan received one for chemistry in 1906. The committee stated that the findings represented the greatest scientific contribution ever made in a doctoral thesis.

To celebrate Marie becoming the first woman in France to get a doctorate, her sister Bronia visited her from Poland. Marie got a new black dress because Bronia insisted she needed one for the celebration. Just like her navy outfit from eight years ago, her new dress meant Marie could work in the lab without worrying about stains.

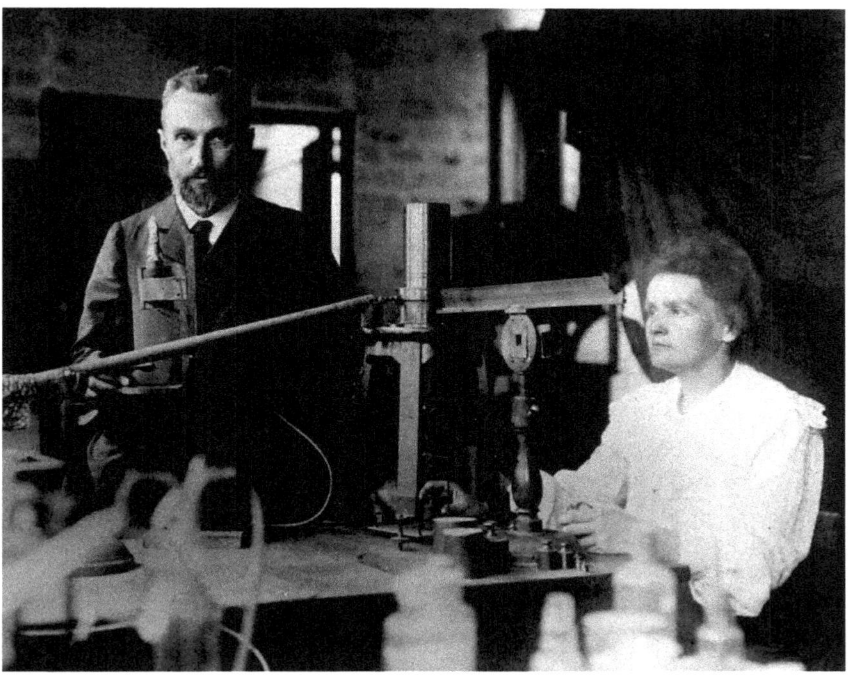

Figure 1. Marie and Pierre Curie in their laboratory. Historical photograph (c. 1904).

## Recognition and a Nobel Prize

The Royal Society of London awarded the Curies one of its top honors, the Davy Medal, in November 1903. Soon after, on 10 December 1903, the Nobel Foundation in Stockholm, Sweden, announced that three French scientists, Becquerel and the Curies, were the joint recipients of the Nobel Prize in Physics for 1903 for their groundbreaking work on radioactivity. The Nobel Committee awarded Becquerel for his discovery of spontaneous radioactivity. Marie and Pierre received the award for their joint research on the radiation phenomena Becquerel had discovered. The discovery of radium and polonium revolutionized chemistry and physics and secured the Curies' place in history. A few months later, the University of Paris appointed Marie as director of research.

In 1905, the French Academy of Sciences elected Pierre as a member. On June 6, 1905, in Stockholm, Sweden, Pierre delivered the Nobel Prize lecture on "Radioactive Substances, Especially Radium," which was in fact their joint address. Pierre finished his speech by talking about how every big scientific step affects us both ways, but he felt "mankind will get more good than bad from the new discoveries."

---

**HISTORICAL SPOTLIGHT**

### The Nobel Prize

Established in 1901 through the will of Alfred Nobel, the Nobel Prize honors individuals whose discoveries or achievements have profoundly advanced humanity. Awarded annually in Physics, Chemistry, Physiology or Medicine, Literature, Peace, and later Economic Sciences, the prize recognizes work that reshapes scientific understanding, cultural expression, or global cooperation.

Each Nobel Prize includes a medal, a diploma, and a substantial cash award, today typically around 10-11 million Swedish kronor (roughly $1 million USD), though the exact amount varies by year. The prize's rigorous selection process and global visibility make it one of the highest distinctions a scientist or scholar can receive. Many landmark breakthroughs, including radioactivity, the structure of DNA, and the discovery of pulsars, were first recognized with Nobel honors, cementing the award's reputation as a symbol of transformative impact.

## Family Life and Hardships

Marie and Pierre had their first daughter, Irène, on September 12, 1897, in Paris. In 1898, they moved to 24 rue de la Glacière, a sober and functional apartment, a stone's throw from the ESPCI. The couple worked tirelessly in the lab, often teaching to supplement their poorly funded research. Pierre's father joined them after the death of his wife, a presence particularly appreciated by Marie, and even more so over the years by little Irène, whom he loved to take care of.

While she was working on her PhD starting in 1900, Marie was also teaching part-time at a college for women, École Normale Supérieure de Sèvres (École normale supérieure de jeunes filles), an institution of higher education to train women teachers for the French secondary system. Six months after Marie received her thesis, they had a second daughter, Ève, on December 6, 1904.

However, Pierre was ill. He was in so much pain, and his legs were trembling, making it hard to stay standing. He consulted a doctor who diagnosed neurasthenia (weakened or depleted nervous system). Pierre's physician treated his neurasthenia with small doses of strychnine, then considered a nerve stimulant capable of counteracting "nervous exhaustion," though it is now known to be a dangerous toxin. While Marie's fingers had cracked and scarred skin, and they both constantly suffered from fatigue. They did not know that radiation could have a detrimental effect on their general state of health. Pierre, who famously claimed radium's radioactivity was a million times that of uranium, would often pull some of it out of his waistcoat to impress his friends. Marie also liked to have a little radium salt by her bed that shone in the darkness.

On Thursday, April 19, 1906, tragedy struck again. On the Rue Dauphine, around 2:30 pm, Pierre crossed the road, passing behind a horse-drawn carriage. In the middle of the road, a fast-moving heavy cart drawn by a horse and Pierre collided. When Pierre tried to grab the horses' harness, it reared, and Pierre slipped on the wet pavement, and the cart pulled by the horse ran over him, fatally fracturing his skull. Pierre's family buried him in the Cimetière de Sceaux, just outside Paris, alongside his mother.

Pierre's death left Marie at 38 years old, a widow with two young daughters, Irène aged 9 and Ève aged 2. Marie remained paralyzed with grief for several days. Marie then decided that she should continue the work they had both started and that this was the best way to honor Pierre's memory. As such, Marie took over Pierre's professorship at the Sorbonne on May 13, 1906, becoming the first woman to hold a chair there.

On November 5, 1906, Marie gave her first lecture after being appointed to Pierre's professorship following his death. It was part of Pierre Curie's physics course, which she took over after his death in April 1906. The hall was full, not only with enrolled students but also with a large crowd from far and near. Journalists and photographers were also in attendance, curious to witness the first female professor in the Sorbonne's history.

In 1907, Marie created a tiny, elite, collaborative school of about 10 children for Irène, Ève, and the children of Paul Langevin, Jean Perrin, and several of her scientific friends. It was sometimes called the Cooperative, the Cooperative School, or informally the "little scientific commune." Marie recruited her closest colleagues, all future scientific giants, to teach the children:

- Marie: Physics and chemistry
- Paul Langevin: Mathematics
- Jean Perrin: Science
- Henri Mouton: Biology
- Édouard Branly: Electricity
- Other Sorbonne colleagues rotated in

These were not ordinary teachers; they were some of the most brilliant scientists in France.

What made the school unique was:

- No rote learning: Everything was hands-on, experimental, exploratory.
- Laboratory work for children: Microscopes, measurements, experiments.
- Outdoor learning: Nature walks, field observations, physical activities.
- Individualized instruction: Each child learned at their own pace.
- Equality and curiosity: No hierarchy, no punishment, no memorization.

It was a radical educational experiment decades ahead of the Montessori or progressive education movements. The children later described it as the most intellectually stimulating period of their lives.

The school ran for about two years, ending around 1909, when the children transitioned into more formal schooling. This small school reflected Marie's deepest values:

- Scientific curiosity
- Equality in education
- Hands-on learning
- Community and collaboration

In 1908, she worked on her late husband's collected writings, and then in 1910, she published the first comprehensive textbook ever written on radioactivity, entitled Traité de radioactivité. The book was a two-volume work that organized the laws, methods, and discoveries of radioactivity into a coherent discipline.

## The French Academy of Sciences (Académie des Sciences)

On 16 November 1910, after *Le Figaro* reported that she was willing to be nominated for election to the French Academy of Sciences (l'Académie des Sciences*)*, many members openly declared that admitting Marie would "set a precedent" they did not want. Her candidacy triggered debates about whether women were "fit" for membership, a barrier no scientific achievement could overcome.

Despite election outcomes sometimes depending on more than just merit, Marie and her top researchers assumed her stellar scientific accomplishments meant she was a guaranteed winner. Yet, it seems it wasn't merit that really mattered in the end. You could see the nasty undercurrents of hate, anti-Semitism, sexism, and anti-science beliefs coming to the forefront in France. Normally, the election was of no interest to the press. Critics claimed that Pierre's research had given her a free ride throughout her career. Some still viewed Marie as a foreign woman in a deeply nationalistic institution. Private correspondence and press commentary repeatedly used her Polish origins against her. She was from Poland, officially Catholic, but her name Sklodowska hinted she might have Jewish roots.

Just seven days before the election, a competing candidate, Édouard Branly, a respected but less groundbreaking scientist whose work on wireless telegraphy the Academy's conservative factions viewed as more aligned, launched a campaign on his behalf. Branly's most influential contribution was the coherer, a simple but sensitive device that detected radio waves and became a foundational component of early wireless communication systems. His experiments helped transform radio from a laboratory curiosity into practical technology, influencing figures such as Oliver Lodge and Guglielmo Marconi. His career reflected the Academy's preference for traditional, applied physics when Curie's groundbreaking work challenged long-standing norms about who could belong in French science.

On January 23, 1911, journalists, photographers, and hordes of curious people observed the vote. The election took place in a tumultuous atmosphere. In the first round, Marie lost by one vote, and in the final by two. As a result, the Academy elected Branly, and Curie was once again denied her membership.

Fifty-eight votes were cast. Even after winning a Nobel Prize in 1903 and getting backing from big names like Jean Perrin, Henri Poincaré, Paul Appell, Gaston Darboux, and the Académie's permanent secretary, they still wouldn't let her in. This event attracted a lot of negative attention worldwide. It deeply wounded Marie and, indeed, Branly, too, himself a well-merited researcher.

---

**SCIENCE SPOTLIGHT**

## The Curie-A Unit Born in 1910

The International Congress of Radiology formally defined the curie (Ci) in 1910 as a standard unit of radioactivity to bring consistency to a rapidly expanding field. The curie represented the activity of one gram of radium-226, the substance whose behavior Marie and Pierre Curie had spent years isolating and measuring. This definition later translated to $3.7 \times 10^{10}$ disintegrations per second, a value that remained central to radiation science for decades.

The unit initially seemed to honor Pierre, but later committees clarified that it acknowledged the work of both Pierre and Marie, whose discoveries made quantitative radioactivity measurement possible. The curie became a foundational tool in medical physics, nuclear research, and radiological safety.

Even though scientists replaced the curie unit with the becquerel in 1975, the name still honors the Curies' pioneering research.

---

## A Public Scandal

Five years after Pierre's death, in late October 1911, Marie's personal life became the subject of public controversy. She had developed a relationship with Paul Langevin, a former student of her husband, who was married.

Langevin had been experiencing marital difficulties for several years and had moved from the family home to a small apartment in Paris. His wife, Jeanne Langevin, was preparing legal action concerning the custody of their four children. During a burglary in Langevin's apartment, thieves stole private letters and later delivered them to the press. When copies of the letters reached Jeanne, she used them in her legal case, but the public leak itself came from the newspapers. The publication of the letters triggered a scathing and deeply misogynistic campaign against Marie in the French press.

The drama culminated on the morning of November 23, 1911, when the newspaper *L'Oeuvre* published extracts from the letters, revealing a deep emotional intimacy between Paul and Marie. Many falsely accused Marie of being a home wrecker and even labeled her Jewish, though she was neither. Angry mobs protested outside her home, and her reputation suffered.

After Langevin suffered repeated insults, he challenged Gustave Téry, the editor of *L'Oeuvre*, to a duel. The duel, with pistols at 25 meters, took place on the morning of November 25, 1911. Finding seconds proved difficult, but he eventually secured the support of mathematician Paul Painlevé, then director of the School of Physics and Chemistry. When the count began unusually fast, Téry never raised his pistol, and Langevin lowered his. No one fired a shot. Journalists later described the quiet field and the pigeons feeding nearby, noting how a moment charged with danger dissolved into a near farce.

Marie stood up in her own defense and forced an apology from the newspaper *Le Temps*. Langevin and his wife settled on December 9, 1911, without Marie's name being mentioned, and the relationship between Paul and Marie ended. During that time, Marie only had a few friends supporting her, namely Jean and Henriette Perrin, Émile and Marguerite Borel, and

André Debierne. Albert Einstein also defended Marie and wrote to her, encouraging her to ignore the slander and focus on her work.

---

**HISTORICAL SPOTLIGHT**

### Einstein's Letter to Curie (1911)

In the last days of 1911, as the stolen letters scandal raged in the Paris press, Albert Einstein wrote a private letter to Marie Curie offering unwavering support. He had just spent time with her at the Solvay Conference, and the attacks on her character appalled him. In warm, direct language, he urged her to ignore "the rabble" and assured her that nothing could diminish the greatness of her scientific work.

Einstein's message arrived before the announcement of her second Nobel Prize, at a moment when the public intensely scrutinized Marie, isolating and exhausting her. His letter became one of the most famous gestures of solidarity in scientific history, an affirmation of her integrity at the very moment it was being publicly questioned.

---

## A Second Nobel Prize

Despite warnings of public protests, Marie remained steadfast and attended the award ceremony in Stockholm on December 10, 1911, as the scandal had nearly impacted her second Nobel Prize.

The Nobel Committee awarded Marie the Nobel Prize in Chemistry for her development of methods to isolate radium and polonium. The citation by the Nobel Committee was, "in recognition of her services to the advancement of chemistry by the discovery of the elements radium and polonium, by the isolation of radium and the study of the nature and compounds of this remarkable element." Chemists considered that the discovery and isolation of radium was the greatest event in chemistry since the discovery of oxygen.

Marie became the first person in history to win Nobel Prizes in two different sciences and remains the only person to have done so in Physics and Chemistry.

However, on December 29, 1911, doctors admitted Marie to a hospital, and they concealed her whereabouts for her safety. After she had recovered somewhat, she traveled to England. Her friend, the physicist Hertha Ayrton, cared for her and ensured the press stayed away. An entire year passed before she could work as she had done before.

## World War I

At the beginning of World War I in 1914, Paris was under threat of invasion by Germany. As a result, the French government had moved to Bordeaux. Once she saw this, Marie sent her 17-year-old daughter Irène and 10-year-old Ève to Brittany to be safe. She then took the 1 gm of radium she had separated and put it in a heavy, 20-kg lead box, moving it to a bank in Bordeaux.

Marie then returned to Paris the next day, where she visited a hospital to see the war wounded. Marie quickly realized the value of having X-rays near the battlefront, giving surgeons the ability to see bullets and shrapnel inside wounded soldiers.

While she understood the physics of X-rays, Marie taught herself their practical application: how to generate them, operate the equipment, develop plates, perform real-time screen examinations, and assemble the components needed for a functioning radiology unit. Bringing X-ray machines to the front meant creating mobile radiology units, and such vehicles simply didn't exist, so Marie set about designing and building them herself.

### Les Petites Curies

In 1914, Marie founded the Red Cross Radiology Service and built her first radiological car from a large Renault. She persuaded wealthy friends and a women's organization to help her fund twenty mobile X-ray units. The team sent these cars to the French front lines to provide surgeons with immediate access to lifesaving imaging. In the war zone, they quickly became known as the "Little Curies" (Petites Curies).

Each car carried a compact X-ray machine powered by a generator or by the vehicle's engine. Inside the darkened compartment, an electric current passed through an X-ray tube, producing penetrating rays that could reveal bullets, shrapnel, and fractures within minutes. Medical staff positioned wounded soldiers on a stretcher or table, placing the injured area between the X-ray tube and a photographic plate. Surgeons often relied on fluoroscopy, a technique in which X-rays passed through the body onto a glowing screen, allowing Marie or her technicians to watch the injury in real time. This immediate visualization helped surgeons locate foreign objects with far greater accuracy than surgery alone.

Working under harsh and dangerous conditions, Marie and her 18-year-old daughter Irène operated several of these units and inevitably encountered radiation exposure. Marie trained more than a hundred women in basic radiological techniques to support the effort. She drove a van herself and assisted surgeons in locating metal fragments, sometimes even giving them quick lessons in elementary geometry to help interpret the images.

The Little Curies transformed wartime medicine. Before their arrival, surgeons often operated "blind," increasing the risk of infection, nerve damage, or amputation. With mobile X-ray imaging, operations became faster, safer, and more precise. Historians estimate that Curie's radiology service benefited hundreds of thousands of wounded soldiers and that surgeons saved tens of thousands of lives and limbs because they could finally see what they were operating on.

Marie's wartime radiology network, with 20 mobile cars and over 200 fixed stations, became a model for modern military medical imaging and cemented her legacy not only as a pioneering scientist but also as a lifesaving innovator.

## Rebuilding After the War

After the Peace Treaty in 1918, Marie could now open her Radium Institute, which the University of Paris and the Pasteur Institute had established jointly and completed in 1914. During the inter-war period, it was France's most well-known research institute internationally. Marie became Director of the Curie Laboratory at the Radium Institute, leading groundbreaking research in physics and chemistry, which quickly became a world center for radioactivity research. Plans for the Radium Institute began in 1909, but construction and wartime delays meant it wasn't fully operational until after World War I.

According to her French biographer Françoise Giroud, she got little backing from the French state. Although the U.S. manufactured radium industrially, Marie could not afford its price. As a result, Marie had to dedicate a significant amount of time to fundraising for her institute. She also became deeply involved when she became a member of the Commission for Intellectual Cooperation of the League of Nations and served as its vice-president for a time. Marie frequently took part in its meetings in Geneva, where she also met the Swedish delegate, Anna Wicksell.

Curie was in her fifties when the war ended. She'd spent all her savings on war bonds and was exhausted. Her dedication was relentless, though.

Her book, *La Radiologie et la guerre (*'Radiology and the War'*)*, came out in 1921. In it, she gave a most informative account of the scientific and human experiences gained in radiology (the use of radiation) during the war. At the war's conclusion, her daughter Irène, a physicist, took a position as an assistant in her mother's laboratory.

## Missy Meloney and the U.S. tour

In 1920, Marie "Missy" Meloney, editor of *The Delineator*, a leading magazine in New York, visited Marie at the Radium Institute to interview her. Marie told Missy that researchers in the U.S. had some 50 grams of radium at their disposal. "And in France, then?" asked Missy. "My laboratory has scarcely more than one gram," was Marie's answer. "But you ought to have all the resources in the world to continue with your research. Someone must see to that," Missy said. "But who?" was Marie's reply in a resigned tone. "The women of America," promised Missy.

After meeting Marie and learning how limited her laboratory resources were, Missy returned to the U.S. determined to help. She launched a nationwide fundraising campaign to buy one gram of radium for Marie's research among women's organizations, civic groups, and readers of *The Delineator*. Only once the effort succeeded did Missy write to Marie with the news: American women had raised $100,000 (equivalent to between $1.58M and $1.67M today). The gram of radium allowed Marie to continue her research and supported her daughters' education.

Marie felt astonished and initially hesitant, worrying that someone might need the money elsewhere. Missy reassured her that donors had given it expressly for her scientific work. Their correspondence transformed the gift from a bold idea into a moment of international solidarity.

Missy also asked Marie to visit the U.S. with her daughters and collect the precious gift in person. Missy had to struggle hard to get Marie to accept a program for her visit on a par with the campaign.

Finally, Missy had to turn to Paul Appell, now the university chancellor, to persuade Marie. Even though she was shy and didn't like being in the spotlight, Marie said yes to going to the U.S. to get a gram of radium from President Warren G. Harding. "I understand that it will be of the greatest value for my Institute," she wrote to Missy.

After it became public in France, the magazine *Je Sais Tout* organized a special event at the Paris Opera. The biggest names in France attended it, including Aristide Briand, the Foreign Minister, who would win the Nobel Peace Prize in 1926. Jean Perrin spoke about Marie's work and the future possibilities her discoveries opened. Famous French actress Sarah Bernhardt read an "Ode to Madame Curie," comparing her to Prometheus's sister. She went from being completely disgraced ten years ago to being adored now.

Finally, Missy had secured a promise from the New York papers that they would not print a single word about the Langevin affair and, almost unbelievably, persuaded editors to withhold every piece of material they had on it so that Marie would feel safe.

Marie received a heroic welcome in New York and visited the Standard Chemical Company near Pittsburgh, which had isolated the radium. This industrial-scale operation mirrored the painstaking work Marie and Pierre had done by hand decades earlier. Because of positive press coverage, Marie became enormously popular in the U.S., and everyone seemed to want to meet her, the great Madame Curie.

In the White House, President Harding presented her with the golden key to the little metal box containing the radium. The radium gift was far more than a symbolic gesture. It solved a critical scientific and financial problem for Marie. Although she had discovered radium, she could not afford even a single gram for her own laboratory; by the 1920s, radium was one of the most expensive substances on earth. The $100,000 raised by American women allowed Marie to continue her research without commercial pressure or dependence on industrial donors.

The gift also marked a rare moment when the public, especially women's organizations, directly supported a woman scientist at the highest level. Presented at the White House, it affirmed Marie's international stature, strengthened scientific ties between France and the U.S., and secured the future of the Radium Institute's medical and chemical research. Her trip in 1921 was an absolute triumph.

## Global Recognition

Marie's groundbreaking discoveries earned her worldwide acclaim. Universities across the United States, the United Kingdom, France, and Poland awarded her honorary degrees in science, medicine, and law. Prestigious scientific bodies, including the Conseil du Physique Solvay in 1911 and the League of Nations' Committee on Intellectual Cooperation in 1922, invited her to serve as a member. Among her many honors were the

1921 Ellen Richards Research Prize, the 1923 Grand Prix du Marquis d'Argenteuil, and the 1931 Cameron Prize from the University of Edinburgh. Although France offered her the nation's highest distinction, the Legion of Honor, Marie declined it, reflecting her humility and her devotion to science over accolades.

Heads of state often welcomed her as a scientific diplomat. Though she rarely spoke publicly on subjects beyond science, she made an exception in May 1933 at the *Congreso Internacional de Cooperación Intelectual,* often translated as The Future of Culture conference, in Madrid. There, she defended the humanistic spirit of scientific inquiry against claims that science had dehumanized modern life. "I am among those who think that science has great beauty," she said, adding that a scientist is "a child placed before natural phenomena which impress him like a fairy tale."

Marie traveled widely to promote radiology, support cancer treatment, and advise governments on scientific and medical policy. Belgium established its own Institut du Radium in 1922, part of a broader European movement shaped by her pioneering work. That same year, she became one of the founding members of the League of Nations' International Commission on Intellectual Cooperation, the body that would later evolve into the United Nations Educational, Scientific, and Cultural Organization (UNESCO), serving alongside Albert Einstein, Henri Bergson, Gilbert Murray, and Jagadish Chandra Bose. Her work emphasized scientific exchange, international collaboration, and the free circulation of knowledge. Her presence lent credibility to scientific and medical initiatives across Europe, and her wartime radiology schools for women became a model for later national training programs.

## Legacy of Institutions and Care

Marie continued her research throughout her life. She established a radioactivity laboratory in Warsaw, Poland, and traveled in 1929 to lay the cornerstone of the Warsaw Radium Institute (*Instytut Radowy*). Bronia directed the Institute when it officially opened in 1932. American supporters funded the Institute with a separate $50,000 gift, which was distinct from the earlier 1921 campaign that had raised $100,000 to purchase a gram of radium for her Paris laboratory.

The Institute was created to advance cancer treatment and radiological research at a time when radium therapy represented the cutting edge of medical science. It combined a hospital with research laboratories dedicated to radiotherapy, medical physics, and radiobiology. Funded through a nationwide public campaign, supported by schoolchildren, workers, and Polish émigré communities, the Institute became one of the most advanced cancer centers in Eastern Europe, offering treatments previously unavailable in Poland.

However, Marie encountered resistance to her leadership at the Radium Institute, even after decades of scientific achievement. Some senior male colleagues were uncomfortable reporting to a woman and attempted to bypass her in administrative decisions. University officials often delayed funding and equipment requests, forcing her to rely on American donors. There were also disputes over hiring and over the allocation of radium, a resource Marie controlled carefully. Although rarely overt, this resistance reflected the lingering discomfort many male scientists felt about a woman directing one of France's most important research centers.

The Marie Curie Hospital in London opened in 1930, after Marie agreed to lend her name to a pioneering 30-bed institution staffed entirely by women and dedicated to treating female cancer patients with radiological methods. Prime Minister Stanley Baldwin led the official opening, with both the Polish and French ambassadors in attendance. The demand for treatment grew so quickly that they created an Outpatients Department in the building's former coal cellar the same year. This allowed women who had undergone radium insertion therapy, mainly for gynecological cancers, to return for regular follow-up care. In 1934, the Duchess of York, later Queen Elizabeth the Queen Mother, opened a new extension equipped with its own X-ray facilities. Renamed the Marie Curie Hospital for Cancer and Allied Diseases in 1938, the institution served patients for several decades before finally closing in 1967.

## Health and Final Years

In 1926, Marie delighted in her daughter Irène's marriage to Frédéric Joliot, the Radium Institute's most talented young assistant. After the wedding, both Irène and Frédéric adopted the hyphenated surname, Joliot-Curie. She soon saw that their relationship would be like her wonderful creative team-up with Pierre. Marie witnessed Irène and her son-in-law Frédéric conduct successful research in the laboratory and their discovery of artificial radioactivity.

Their past work extracting radium exposed Marie and Pierre to dangerous levels of radiation, although at the time they were unaware of it. Years of exposure took a toll. Marie experienced physical exhaustion, raw hands, and other symptoms of radiation sickness. By the late 1920s, Marie's health had declined, and she suffered from cataracts and weakened immunity. Marie took several restorative trips to the mountain and spa regions, especially in the French Alps (Haute-Savoie area) and the spa towns in the Pyrenees, famous for their mineral baths and therapeutic climates.

In 1934, Marie retired to the Sancellemoz Sanatorium in Passy, Haute-Savoie, France, with her daughter Ève for care. Sancellemoz was a high-altitude sanatorium originally built for tuberculosis patients, later used for rest cures and convalescence.

## Death and Honor

On July 4, 1934, at 66 years of age, Marie died from radiation-induced aplastic anemia, which is a failure of the bone marrow caused by many years of exposure to high-energy radiation. Her daughters Irène and Ève attended her funeral together with her son-in-law Frédéric, leading scientists such as Paul Langevin and Jean Perrin, and many family members, friends, and colleagues from the scientific community. For 61 years, she rested in the Curie family burial plot at the Cimetière de Sceaux, beside Pierre.

In 1995, French President François Mitterrand ordered the transfer of Marie and Pierre Curie's ashes from the Cimetière de Sceaux to the Panthéon in

Paris to honor her scientific achievements, correct the historical exclusion of women from the nation's highest tribute, and reunite her with Pierre as symbols of France's intellectual heritage.

The Panthéon is a monumental mausoleum in the Latin Quarter of Paris, overlooking the old university district. Initially, the Panthéon was a church dedicated to Saint Genevieve, but the French Revolution transformed it into a place honoring France's greatest citizens. It stands on the Montagne Sainte-Geneviève, in the heart of the academic district where Marie lived when she first arrived in Paris.

The Panthéon holds the remains of many major French figures, including Voltaire, Victor Hugo, Jean Moulin, and Louis Braille. Marie was the first woman buried in the Panthéon on her own merit. For generations, "great men" had exclusively occupied it, and she was the first woman they honored for her own achievements rather than as someone's wife. Women fighters in the French Resistance, Germaine Tillion, Geneviève de Gaulle-Anthonioz, as well as Holocaust survivor, politician, and feminist Simone Veil, and Civil Rights activist, Résistante, and singer Josephine Baker, would later follow her.

Her laboratory notebooks were radioactive, but Marie's remains were not dangerously contaminated. When they transferred her and Pierre to the Panthéon, radiological tests confirmed they did not need special shielding, and they constructed her tomb of ordinary stone like the others in the crypt. Marie received a full state ceremony for her reinterment, which was attended by French President François Mitterrand, Polish President Lech Wałęsa, Ève Curie, members of the Joliot-Curie family, Danielle Mitterrand, Prime Minister Édouard Balladur, and other political and scientific dignitaries.

---

**HISTORICAL SPOTLIGHT**

## Irène and Ève Curie and Their Achievements

Irène Joliot-Curie, the elder daughter of Marie and Pierre Curie, became one of the most influential scientists of the twentieth century. Trained in the laboratory her parents built, she carried their work into a new era. She and

Frédéric Joliot, her husband, shared the 1935 Nobel Prize in Chemistry for discovering artificial radioactivity, paving the way for today's nuclear science and medicine. Irène went on to serve as a professor at the Sorbonne and later as the director of the Radium Institute, shaping France's early nuclear research programs. Beyond her scientific achievements, she was a committed advocate for women in science and an active participant in the French Resistance during the Second World War, helping to safeguard scientific materials and support underground networks.

Ève Curie, the younger daughter, chose a very different path. Gifted in languages, music, and writing, she became an accomplished author and journalist. Her biography of her mother, Madame Curie, published in 1937, became an international bestseller and won the National Book Award in the United States. During the Second World War, Ève worked as a war correspondent, reporting from Africa, Asia, and Europe with clarity and courage. She later joined the Free French Forces and, after the war, dedicated much of her life to humanitarian work. As a senior official at UNICEF, she promoted global health, children's welfare, and international cooperation.

The Curie legacy continued into the next generation. In 1948, Michel, Paul Langevin's grandson, married Hélène Joliot Curie, Marie's granddaughter. This symbolically united two families. Controversy once intertwined with their histories. Hélène Langevin-Joliot became a distinguished nuclear physicist in her own right. Later in life, she undertook a careful study of Marie and Pierre's laboratory notebooks, tracing their experimental methods and collaborative process with a scientist's eye and a granddaughter's insight. Her work helped illuminate how the Curies thought, reasoned, and worked together during the foundational years of radioactivity research.

Together, Irène, Ève, and Hélène carried forward the Curie spirit, one through scientific discovery, one through writing and humanitarian service, and one through preserving and interpreting the scientific legacy that shaped the modern age.

## Legacy

Marie's last book, *Radioactivité,* which she was working on at the end of her life, was unfinished at the time of her death in July 1934. The manuscript brought together decades of research on radioactive substances, tracing the evolution of the field she helped create. Marie worked on the book while her health was declining, determined to leave a clear, rigorous account of the principles and methods that shaped modern nuclear science. After her death, her daughter Irène and son-in-law Frédéric prepared the text for publication, preserving Marie's voice and scientific precision. They published it in 1935, making it a posthumous publication.

An issue later faced was that the papers Marie and Pierre left behind still give off significant radioactivity. Anyone who wants to consult the three black notebooks at the Bibliothèque Nationale de France, containing their work from December 1897 and the following three years, must sign a radiation-risk consent form. People will have to do this for a very long time, as radium-226 has a half-life of about 1,600 years.

Marie and Pierre's work has since led to the development of techniques to isolate radioactive isotopes and study their properties, and the use of radium in cancer treatment and medical research.

Marie's legacy is monumental:

- In 1944, scientists named the 96th element curium in honor of Marie and Pierre.
- A 2009 poll revealed people voted her the most inspirational woman in science.
- The UN declared 2011 the International Year of Chemistry, while Poland and France celebrated the Year of Marie Curie.

Through grit, curiosity, and relentless work, Marie transformed science. Marie broke barriers for women in science, defied societal expectations, becoming the first female professor at the Sorbonne and a global icon of research. She showed the world that passion, perseverance, and curiosity can change history.

Albert Einstein once wrote to her, "Marie Curie is, of all celebrated beings, the only one whom fame has not corrupted."

## Modern Imaging and Radiation Therapy

The tools used in hospitals today trace their lineage directly to the discoveries Marie helped set in motion. The same principles of radioactive decay, energy release, and atomic transformation that she first explored in her Paris laboratory are now relied upon by modern imaging and cancer treatment technologies.

Although people often mention CT (Computed Tomography) and PET (Positron Emission Tomography) scans together, they show distinctly different types of information. CT scans use rotating X-rays to snap detailed images of your body's insides, like bones and organs, similar to a 3D X-ray. A PET scan, on the other hand, shows how things work instead of what they look like. Patients receive a tiny amount of a radioactive tracer, which emits positrons as it decays. Areas of high metabolic activity, such as tumors, light up on the scan. Combining CT and PET images allows physicians to see both anatomy and underlying biological activity, providing a far more complete picture of disease.

A related technique, SPECT imaging (Single Photon Emission Computed Tomography), also uses radioactive tracers but detects gamma rays instead of positrons. SPECT is especially useful for studying blood flow, heart function, and certain neurological conditions. Together, PET and SPECT form the backbone of modern nuclear medicine, fields that would not exist without the Curies' early work on radioactive elements.

Radiation remains one of the most powerful tools for treating cancer. Linear accelerators (LINACs) use high-energy beams of radiation to destroy cancer cells and spare surrounding tissue, shaping and targeting these beams with extraordinary precision. Another approach, brachytherapy, places tiny radioactive sources directly inside or next to a tumor. Because the radiation acts from within, the dose is intense but highly localized, making it especially effective for prostate, cervical, and certain breast cancers.

Some tumors require even more specialized techniques. The Gamma Knife, despite its name, is not a knife at all but a device that focuses hundreds of narrow gamma-ray beams on a single point inside the brain. Each beam is too weak to damage healthy tissue on its own, but where they converge, the combined dose is strong enough to destroy a tumor with remarkable accuracy.

Scientists are developing targeted radiopharmaceutical therapy, a new frontier where they attach radioactive atoms to molecules that seek out cancer cells. One example is lutetium-177, a radioisotope that delivers radiation directly to tumors while minimizing exposure to the rest of the body. These treatments represent a marriage of chemistry, biology, and nuclear physics, disciplines that Marie helped shape more than a century ago.

The technologies of today, from diagnostic imaging to precision cancer therapy, continue to build on the principles she uncovered: that atoms can transform, that we can measure their emissions, and that we can harness radioactivity to heal once we understand it.

## Why She Matters Today

Marie's discoveries continue to shape modern medicine and save lives every day. The principles she uncovered, radioactivity, radioactive decay, and the behavior of unstable atoms, form the backbone of several essential medical technologies.

Radiotherapy is one of the most widely used cancer treatments in the world, and it exists because of Marie's work. The first radium treatments began only a few years after her discovery, and the basic idea remains the same today: targeted radiation damages the DNA of cancer cells so they can no longer divide. Modern radiotherapy uses safer, precisely controlled sources rather than radium, but the underlying science traces directly to Marie's laboratory. Radiation therapy helps millions of patients each year, stemming from her insistence that radioactivity could heal as well as be used for research.

Positron Emission Tomography (PET) scans rely on radioactive tracers that emit positrons, particles whose behavior was first understood through the study of radioactive decay. PET imaging allows physicians to see metabolic activity inside the body, revealing tumors, heart disease, and neurological disorders long before structural changes appear. Every PET scan is a direct descendant of Curie's insight that atoms themselves can emit measurable, characteristic radiation.

Nuclear medicine uses small amounts of radioactive isotopes to diagnose and treat diseases. Whether it is iodine-131 for thyroid disorders, technetium-99m for imaging, or targeted radiopharmaceuticals for cancer, the field rests on Marie's foundational work: isolating radioactive elements, measuring their emissions, and understanding their transformations. Her methods made it possible to identify isotopes, track them inside the body, and harness them safely for medical use.

Marie's own illness, radiation-induced aplastic anemia, became a powerful lesson in the need for protection. Today's radiation safety standards, from shielding and exposure limits to dosimetry badges, exist because early researchers like Marie unknowingly absorbed dangerous doses. Decades of study into how radiation interacts with tissue, much of it sparked by the very discoveries she made, form the basis of modern protocols. Her life's work not only opened new scientific frontiers but also taught the world how to handle radiation responsibly.

Today, Marie remains one of history's greatest scientists. Her discoveries transformed physics, chemistry, and medicine, while her determination showed that curiosity, perseverance, and courage can overcome even the greatest barriers. More than a century later, hospitals, research laboratories, and universities around the world continue to build upon the foundations she helped create, inspiring generations of scientists to explore the unknown.

# Timeline

| Year | Event |
|---|---|
| 1867 | Born in Warsaw |
| 1891 | Arrived in Paris and enrolled at the Sorbonne |
| 1893 | Earned Licence ès Sciences Physiques |
| 1894 | Earned Licence ès Sciences Mathématiques |
| 1895 | Married Pierre |
| 1897 | Birth of Irène Curie |
| 1898 | Discovered polonium and radium |
| 1902 | Isolated pure radium salts |
| 1903 | Awarded a Docteur ès Sciences Physiques; Awarded the Nobel Prize in Physics |
| 1904 | Birth of Ève Curie |
| 1906 | First female professor at the Sorbonne |
| | Death of Pierre |
| 1910 | Published *Treatise on Radioactivity* |
| 1911 | Awarded the Nobel Prize in Chemistry |
| 1912 | Construction began at the Radium Institute |
| 1914-1918 | Developed the Little Curies |
| 1921 | U.S. fundraising tour for radium |
| 1922 | Appointed to the League of Nations' Committee on Intellectual Cooperation |
| 1932 | Opened the Radium Institute |
| 1934 | Died at 66 |
| 1995 | Interred in the Panthéon |

"I was taught that the way of progress was neither swift nor easy."

— *Marie Curie*

# JANE GOODALL

(1934-2025)

## Primatologist and Anthropologist

Transformed our understanding of chimpanzees through groundbreaking field research.

Original portrait illustration by Osmo, based on a historical reference photograph (c. 1986)

"Only if we understand, can we care. Only if we care, will we help.
Only if we help, shall they be saved."
—*Jane Goodall*

Jane Goodall began with little more than curiosity, patience, and an unwavering love for the natural world. When society rarely encouraged women to pursue a career in science, Jane followed her instincts into the forests of Gombe. Her observations would reshape our understanding of animals and of ourselves.

## Childhood

Valerie Jane Morris-Goodall was born on April 3, 1934, in Hampstead, London. She was the eldest daughter of Mortimer Herbert Morris-Goodall, a businessman and engineer who also raced cars, and Margaret Myfanwe Joseph, known as Vanne, a secretary who later became a successful writer. Four years later, on the same date, April 3, 1938, her sister Judith was born.

From an early age, Jane showed an intense curiosity about the natural world. Encouraged by her mother, Vanne, she spent hours observing animals, asking questions, and recording what she saw. She studied earthworms to understand how they moved without legs, watched birds and insects in the garden, and carefully noted their behavior.

One childhood moment became especially memorable. At four years old, Jane disappeared for several hours, later found quietly sitting in a henhouse, determined to learn how chickens lay eggs. Rather than scolding her, her mother praised her patience and curiosity, an approach that would shape Jane's confidence and lifelong approach to learning.

Animals were a constant presence in her life. She grew up with pets, including a dog named Rusty, a pony, and a tortoise. But one gift stood out above all others: a stuffed chimpanzee named Jubilee, given to her by her father when she was a baby. Inspired by a real chimpanzee born at the London Zoo, Jubilee became a treasured companion that Jane kept throughout her life, despite Jane saying, "My mother's friends were horrified by this toy, thinking it would frighten me and give me nightmares."

As she grew older, books fueled her imagination. Jane read Tarzan of the Apes and Doctor Dolittle, becoming fascinated with Africa and its wildlife. She dreamed of traveling there, living among animals, and studying them in their natural environment. She later joked that "Tarzan had married the wrong Jane."

By the age of ten, she had already decided on her future: she would go to Africa, study animals, and write about them. At a time when such ambitions were unusual, especially for a girl, her mother offered steady encouragement, telling her that if she truly wanted something, she must work hard, take opportunities, and never give up.

## Wartime Years and Education

When World War II began, Jane, Vanne, and her sister moved to Bournemouth, where they lived with relatives in a large house called The Birches. This became her childhood home in England.

Jane attended Uplands School, an independent school close to her home in nearby Poole, earning her school certificate in 1950 and a higher certificate in 1952. Although she was academically capable, university was not an option because of financial constraints.

Instead, following Vanne's practical advice that, "Well, if you are set on going to Africa or some other foreign place, if you learn secretarial work, then you can get a job anywhere in the world," Jane trained as a secretary. In 1953, aged 19, Jane enrolled at Queen's Secretarial College in London. She learned to type, shorthand, and bookkeeping skills that would later prove crucial.

After finishing school, Jane took a series of jobs, including work as a waitress, a secretary at the University of Oxford, and a position with a documentary film company in London.

In 1956, a school friend called Clo invited Jane to visit her at her family's farm in Kenya. Jane was excited and determined to go, so to save enough money to make her first trip to Africa, she spent the summer of 1956 working long hours as a waitress in a local hotel in Bournemouth, living at home to save money to fund her trip.

This trip would be a pivotal moment in her life and would mark the beginning of the life she had dreamed of since childhood.

## Meeting Louis Leakey

In 1957, at 23 years old, Jane left London aboard the *Kenya Castle*, a 576ft (176m) steamship capable of carrying over 500 passengers. After a three-week journey, with stops in Italy and Egypt, Jane arrived in Mombasa, Kenya, on April 2, 1957.

Jane traveled to the farm of Clo, in South Kinangop in the White Highlands, near Nairobi. This region was part of British Kenya, then a colony and protectorate of the United Kingdom. On the journey, Jane saw her first giraffe up close, an experience that left a lasting impression. She spent several weeks on the farm before moving on to Nairobi.

While in Kenya, Jane found work as a secretary. Through connections that she made there, Jane's contacts encouraged her to contact Dr. Louis Seymour Bazett Leakey, a renowned archaeologist and paleontologist, because of her strong interest in animals. Within weeks of arriving in Kenya, she secured a meeting with him.

Leakey was curator of the Coryndon Museum (now the Kenya National Museum). Born in Kenya in 1903, he had spent part of his education in England but kept a deep connection to Africa. During their first meeting, Leakey showed Jane around the museum and questioned her extensively about the animals on display. Her knowledge, curiosity, and observational abilities impressed him.

Leakey's work extended beyond fossils. He believed that studying the behavior of living primates could help unlock the story of early evolution. Chimpanzees remained poorly understood, even though they share approximately 98.8% of their DNA with humans. Unbeknownst to Jane, Leakey was actively searching for someone to study them in the wild.

As an anthropologist, the fieldwork methods pioneered by Bronislaw Malinowski during his fieldwork in the Trobriand Islands influenced Leakey starting in 1914. Malinowski emphasized immersive, long-term observation without preconceived theories and believed many formally trained scientists lost objectivity through rigid academic frameworks. Instead, he valued what would now be called a "grounded" approach: careful, unbiased observation.

Although Jane had approached him simply to discuss animals, Leakey was so impressed by Jane's knowledge that he offered her a position as his secretary at the natural history museum. Coincidentally, the role had become available just two days earlier. However, Leakey had larger plans in mind, but before proceeding, he wanted his wife, Mary Leakey, a respected paleoanthropologist, to meet Jane.

Mary Leakey tested Jane's instincts in an unexpected way. During a Pony Club outing, she invited Jane to ride her horse, Sherry, known for behaving badly and walking backwards with unfamiliar riders. When the horse began acting strangely, Jane dismounted and removed the saddle, discovering two painful saddle sores underneath. Her sensitivity to the animal's discomfort immediately impressed Mary Leakey, who gave her approval.

---

**HISTORICAL SPOTLIGHT**

## Louis Leakey (1903–1972)

Louis Seymour Bazett Leakey was born in Kabete, British East Africa, to Church of England missionaries working among the Kikuyu. He spent his early years in a thatched hut surrounded by wildlife and local children, grew up speaking Kikuyu as fluently as English, and was formally initiated into the Kikuyu community. This deep immersion in East African landscapes and cultures shaped his lifelong fascination with human origins.

Leakey studied archaeology and anthropology at St John's College, Cambridge, earning a double first. His personal life was complex: he married Frida Avern in 1928 and had two children, but the marriage ended after he began working closely with illustrator Mary Nicol. He and Frida divorced in 1936, and he married Mary the same year. Their scientific partnership became one of the most influential in 20th-century paleoanthropology.

Together, Louis and Mary conducted groundbreaking excavations at Olduvai Gorge, producing discoveries that helped establish Africa as the cradle of humankind. Louis also transformed primatology by mentoring Jane Goodall, Dian Fossey, and Biruté Galdikas, the "Trimates" whose long-term studies reshaped our understanding of great apes. He spent his final years lecturing, fundraising, and supporting research across the East African Rift. Louis died in London in 1972. Kenya became his final resting place near his parents.

---

**HISTORICAL SPOTLIGHT**

## Mary Leakey (1913–1996)

Mary Douglas Leakey, born Mary Nicol in London in 1913, spent her childhood traveling across Europe, especially the Dordogne, where prehistoric caves first sparked her fascination with archaeology. Independent and strong-willed, she disliked formal schooling but excelled in fieldwork, sketching stone tools and learning geology and archaeology through direct experience.

In 1934, she met Louis Leakey, who hired her to illustrate Adam's Ancestors. Their partnership quickly deepened; they married in 1936 and later raised three sons: Jonathan, Richard, and Philip. Moving to Kenya, Mary introduced more rigorous excavation methods to East Africa and began the work that defined her career.

Her discoveries at Olduvai Gorge transformed the study of human origins. In 1959, she uncovered the skull of Paranthropus boisei ("Zinj"), followed by early Homo habilis remains, showing that multiple hominin species once coexisted in East Africa. She later led the Laetoli excavations, where in 1976 her team revealed the famous hominin footprints preserved in volcanic ash, the earliest clear evidence of upright walking.

After Louis died in 1972, Mary continued lecturing, publishing, and supporting research. She retired to Nairobi in 1983 and worked on her manuscripts until she died in 1996. Her legacy endures in the methods she refined, the discoveries she made, and the generations of scientists she inspired.

## Olduvai Gorge

Soon after, Leakey invited Jane to join him and Mary Leakey on an archaeological expedition to Olduvai Gorge in Tanganyika (now Tanzania), one of the world's most important fossil sites. The site would later play a key role in proving that human origins lie in Africa. Jane accepted the offer and traveled over 300 miles from Nairobi to the remote site.

At Olduvai Gorge, Jane adapted quickly to the demanding conditions. She rose at dawn, worked through the morning heat, rested during the hottest part of the day, and resumed work in the evening. Her tasks included carefully excavating and cataloging fossilized bones and work that required patience and precision.

Leakey was still searching for the right person to study chimpanzees in the wild. He began to consider that women could be particularly well-suited to the task, believing they often showed greater patience and observational focus than men. Jane stood out. Her passion for animals, her resilience, and her ability to observe without bias made her an ideal candidate.

Importantly, Leakey also valued her lack of formal scientific training. Prevailing theories had not yet shaped her, he believed, which allowed her to approach the work with a fresh perspective and objectivity.

He proposed that Jane travel to Gombe Stream, a remote region on the shores of Lake Tanganyika, to study wild chimpanzees and explore potential links between their behavior and human evolution. He warned her that the work would be difficult and potentially dangerous. The terrain was rugged, the animals unpredictable, and the chimpanzees themselves far stronger than humans.

Previous attempts to study chimpanzees had largely failed. Large research teams often frightened the animals, while shorter expeditions did not allow enough time to gain meaningful insight. Leakey emphasized that success would require patience, endurance, and a long-term commitment. For Jane, this was the opportunity she had dreamed of since childhood, so she accepted.

Later, Leakey picked three women: Jane for chimpanzees, Dian Fossey for gorillas, and Birutė Galdikas for orangutans. They would become known as

"Leakey's Angels" or the "Trimates," revolutionizing the study of primates and reshaping our understanding of humanity's closest relatives.

---

**HISTORICAL SPOTLIGHT**

## Olduvai Gorge: The Cradle of Human Evolution

In northern Tanzania, Olduvai Gorge is a deep ravine that was carved into the eastern Serengeti Plain, which is now part of the Ngorongoro Conservation Area. Stretching roughly 30 miles (48 km) and plunging 90 meters deep, it exposes a continuous sequence of geological layers spanning 2.1 million to 15,000 years. These strata preserve one of the richest records of early human evolution ever discovered.

Here, researchers have recovered over 60 hominin fossils, along with the longest known archaeological sequence of early stone-tool industries. In 1959, Mary Leakey uncovered the famous Paranthropus boisei skull ("Nutcracker Man"), dated to about 1.75 million years ago. Soon after, the Leakey team identified early Homo habilis remains, including the reconstructed skull known as "Twiggy." These findings helped confirm Africa as the birthplace of humankind.

Olduvai Gorge's layered beds, from Bed I through the Naisiusiu Beds, reveal shifting lakes, volcanic eruptions, and ancient river systems that preserved tools, animal bones, and hominin remains in remarkable detail. Designated a UNESCO World Heritage Site in 1979, Olduvai Gorge remains a cornerstone of paleoanthropology and a window into the deep past of our species.

---

---

**HISTORICAL SPOTLIGHT**

## Leakey's Angels

In the 1960s and 1970s, Louis Leakey championed a bold idea: that long-term studies of great apes could illuminate the deep evolutionary roots of human behavior. To pursue this vision, he selected and funded three women, later nicknamed "Leakey's Angels," whose patience, observational skills, and independence made them uniquely suited to the work.

Jane Goodall began her research at Gombe Stream in 1960, documenting chimpanzee tool use, hunting, and rich emotional lives. Her discoveries challenged long-held beliefs about what distinguished humans from other animals.

Dian Fossey established the Karisoke Research Center in Rwanda in 1967, where her close studies of mountain gorillas revealed their gentle social bonds and helped galvanize global conservation efforts.

Biruté Galdikas launched her fieldwork in Borneo in 1971, becoming the foremost expert on orangutans and showing how their slow, solitary lives make them especially vulnerable to habitat loss.

Together, these three scientists reshaped primatology. Guided by Leakey's belief in immersive, long-term observation, they produced discoveries that continue to influence anthropology, conservation, and our understanding of what it means to be human.

## Training in London

After Jane agreed to Leakey's proposal, he began working to secure funding for the project. In the meantime, Jane returned to England in December 1958 and moved to London. There, she worked at Granada Television's film library while spending her spare time at the London Zoo observing primates and studying their behavior.

Leakey also arranged for Jane to receive informal training from leading experts. She met Dr. Osman Hill, a respected primatologist and anatomist based at the Royal Free Hospital School of Medicine, and John Napier, an anatomist and specialist in primate hands and locomotion at St. Bartholomew's Hospital Medical College.

Under Hill's guidance, Jane learned how to observe and record primate behavior systematically, gaining an understanding of comparative methods used in scientific study. Napier introduced her to primate anatomy, including skeletal structure, musculature, and movement. This combination of behavioral observation and anatomical study gave Jane a practical foundation for her future fieldwork.

Through this training, Jane realized the intelligence and complexity of chimpanzees. It reinforced her belief that studying them in their natural environment would be both important and transformative.

Not everyone supported Leakey's decision to send Jane. Many experts objected to her lack of formal scientific education and the absence of a university degree. Despite this, Leakey persisted. He eventually secured six months of funding from the National Geographic Society and got permission for Jane to work in the Gombe Stream Game Reserve in Tanganyika, then under British administration.

There was, however, one further obstacle. It was considered unsafe for a young, unmarried English woman to travel alone into a remote African reserve. David Anstey, the chief warden, insisted that Jane have an escort.

Jane turned to the person who had always supported her ambitions: her mother. At 54 years old, Vanne agreed to join her daughter for the first four months of the expedition.

## The Vervet Monkey Excursion

In 1960, Jane set out on her second and historic journey to Africa. She and Vanne flew from London to Nairobi, which served as the primary hub for research in East Africa and where Leakey was based.

However, delays prevented their journey to Gombe Stream. Local conflicts among fishermen in the region made travel unsafe. Rather than lose valuable time, Leakey arranged for Jane to gain field experience elsewhere.

Jane and Vanne traveled to Lake Victoria, Africa's largest lake, which borders Tanzania, Uganda, and Kenya. There, Jane began observing vervet monkeys living on an island in the lake.

Over ten days, she patiently waited to gain the animal's trust, gradually observing them at closer range. The experience provided her with valuable early fieldwork skills, teaching her how to watch the monkey troop without disturbing them and how to interpret their behavior in a natural setting.

However, an unsettling encounter with a hostile crocodile poacher on the island cut her time there short. The incident prompted Jane and Vanne to leave the area sooner than planned.

Despite the interruption, the experience had given Jane a first real taste of field research, an important step before beginning her work with chimpanzees.

---

**SCIENCE SPOTLIGHT**

### Vervet Monkeys (Chlorocebus pygerythrus)

Vervet monkeys are highly social African primates that live in troops and communicate through a range of vocalizations and behaviors. Before beginning her chimpanzee research, Jane spent several days observing vervets on an island in Lake Victoria. The experience taught her many of the field skills she would later use at Gombe, including patient observation, interpreting animal behavior, and gaining the trust of wild primates. Vervets remain important to science today because their complex social interactions and alarm-call system provide insights into communication and behavior among primates.

---

## Kigoma

As Jane and Vanne traveled southwest from Lake Victoria to Gombe Stream, they experienced a temporary delay in the port town of Kigoma, Tanzania, on the northeastern shores of Lake Tanganyika, about 16 miles south of their destination.

The region was in a state of unrest. Back on June 30, 1960, the Belgian Congo broke away from Belgium and became the Democratic Republic of the Congo. Within days, however, the country descended into crisis as parts of the army rebelled and violence spread across several provinces. Refugees began crossing into neighboring countries, and Kigoma, near the borders of both the Democratic Republic of the Congo and Burundi, quickly filled with people seeking safety.

Authorities restricted travel in the area while they assessed the situation. Jane and Vanne, along with their cook, Dominic, had to wait until conditions stabilized.

Eventually, they could continue. On July 14, 1960, at 26 years old, Jane and her companions arrived at Gombe Stream to begin her study of wild chimpanzees.

## Gombe Stream Game Reserve

Gombe Stream Game Reserve is on the eastern shore of Lake Tanganyika in what is now Tanzania. Though small, covering just 14 square miles (35 square kilometers), it is one of the most biologically rich regions in Africa.

Varied and demanding, the landscape features steep valleys, dense tropical forests, rocky hills, and thick canopy cover within a hot, humid climate. Despite its size, Gombe Stream Game Reserve is home to a remarkable concentration of wildlife, particularly primates, including chimpanzees, baboons, red colobus monkeys, vervet monkeys, and red-tailed monkeys. The area also supports hundreds of bird species, along with animals such as bush pigs, hippopotamuses, various snakes, small antelopes, and leopards. Leopards are one of the primary predators of chimpanzees.

On 14 July, Jane arrived by boat with only basic equipment: a notebook and pen, binoculars, and a whistle in case of danger. A game warden escorted Jane and her group to a small campsite in the forest.

The camp was simple and remote. Jane and Vanne used the same tent, which had cots, a table, and chairs inside. Dominic, their cook, lived nearby and helped maintain daily life, preparing meals and supporting the small research operation. Other local assistants, including a camp helper named Saidu, helped carry supplies and set up the site.

The isolation was striking. The warden who escorted them doubted they would last even two weeks. Jane believed the study might take three years. In the end, she would remain for over two decades.

In the early days, even the local game scouts were wary. Their camp, along the lakeshore, did not initially welcome Jane and Vanne. There was suspicion that they might be British government agents, sent to justify expanding protected land by reporting higher numbers of chimpanzees. It took several weeks for trust to build, and only then could the actual work begin.

## Starting Out

On her first day, Jane saw baboons and a red-gold bushbuck. The chimpanzees, however, remained elusive. Although she knew they were there, they kept their distance, fleeing at the first sight of her and disappearing into the dense forest before she could observe them.

The first few weeks were discouraging. Jane worried constantly that time and funding would run out before she made any actual progress. She feared Leakey might lose faith in her.

During this time, Vanne played a quiet but essential role in the camp's survival. With basic nursing experience from her wartime service, she quickly became known among local villagers for her calm and capable manner. People began arriving at the camp seeking help with cuts, infections, burns, fevers, and stomach illnesses. With only a small medical kit, she cleaned wounds, dressed injuries, and offered simple remedies. Her presence helped build trust between the small research team and the surrounding community, something that would prove invaluable in the months ahead.

During the rainy season, Jane fell seriously ill with what was likely malaria. She became weak and bedridden, unable to climb the steep hills to search for chimpanzees. For a time, it seemed the project might end before it had truly begun. Vanne also fell ill, though less severely. Despite this, Vanne continued to care for Jane, assist the camp, and support visiting villagers. Her steady presence kept the camp functioning during a critical period.

Vanne remained at Gombe for five months, leaving in late November 1960. By then, Jane was finally well enough to resume work.

---

## SCIENCE SPOTLIGHT

### Ethology: The Science of Animal Behavior

Ethology is the study of animal behavior, a field that took shape in the early 1900s through the work of Konrad Lorenz, Ivan Pavlov, B. F. Skinner, and John B. Watson. For centuries, many scientists saw animals as machines guided by instinct. This perspective stemmed from René Descartes' argument that animals lacked minds and existed solely for human purposes.

Early ethologists depended heavily on controlled laboratory experiments and often dismissed observing animals in the wild as unscientific. Jane Goodall challenged this assumption. By living among the chimpanzees of Gombe, following them through forests, and earning their trust over months and years, she achieved a level of access no scientist had reached before.

Her observations overturned long-held beliefs. Chimpanzees proved to be intelligent, emotional, and deeply social, forming lasting bonds, expressing affection and grief, and navigating complex hierarchies. Jane's use of terms like "childhood," "motivation," and "mood" drew criticism for anthropomorphism, but her meticulous field notes supported her interpretations.

Her work also revived Charles Darwin's argument in The Expression of the Emotions in Man and Animals that humans and other animals share emotional expression. Jane offered strong evidence supporting this view and corrected major misconceptions by documenting that chimpanzees hunt and consume meat and that their societies possess structure, dynamism, and are much closer to human communities than previously thought.

## Getting to Know the Chimpanzees of Gombe

The chimpanzees Jane set out to study belonged to the Kasekela chimpanzee community, a group of wild eastern chimpanzees living near Lake Tanganyika.

During the early months, things were slow and difficult. The chimpanzees would flee whenever she approached. Jane often could not get closer than 500 yards before they vanished into the forest. For weeks, she saw little more than brief glimpses of dark shapes moving through the dense foliage.

Many researchers might have given up, but Jane did not. Instead, she changed her approach. Rather than pursuing the chimpanzees, she slowed down. She climbed to high vantage points and spent hours scanning the hillside. With the help of local Tanzanian assistants, who tracked chimpanzees, maintained the camp, and supported daily logistics, she began to understand their movements.

Gradually, she established a routine. Each day, she returned to the same area overlooking a feeding site in the Kakombe Valley. She arrived quietly at the same time each morning, allowing the chimpanzees to grow accustomed to her presence without feeling threatened.

Slowly, their behavior changed. After four months, Jane experienced a breakthrough. While walking through the forest, she encountered a large male chimpanzee calmly foraging. Unlike the others, he did not run. Jane named him David Greybeard after the distinctive silver hair on his chin.

David Greybeard became the first chimpanzee to accept her presence. Through him, Jane began to gain the trust of others in the group. Over time, the chimpanzees grew more tolerant. Within a year, Jane could approach the chimpanzees within 30 feet. After two years of daily contact, they no longer showed fear and would sometimes approach her directly.

This access allowed Jane to observe behaviors never documented in the wild. She came to view chimpanzees not as a single species, but as individuals with distinct personalities, social relationships, emotions, and long-term bonds.

To deepen her observations, Jane introduced a controlled feeding system she later referred to as the "banana club." By providing small amounts of food at a consistent location, she could bring chimpanzees together and study their interactions. This approach allowed her to identify individuals, track relationships, and study behavior in greater detail. It also sped up the process of habituation, helping the chimpanzees become comfortable with her presence.

Through careful, sustained observation, Jane documented patterns of communication, including a range of vocalizations and gestures. She identified over twenty distinct sounds, suggesting a primitive but meaningful system of communication.

Her work was beginning to change how scientists understood chimpanzees and, by extension, how they understood us.

---

**SCIENCE SPOTLIGHT**

## What Is a Chimpanzee?

Chimpanzees (Pan troglodytes) are great apes native to the forests and savannas of equatorial Africa. They live in fluid, dynamic communities that may include a few dozen to more than a hundred individuals, forming intricate social networks built on alliances, hierarchies, and long-term bonds.

Humans and chimpanzees share a common ancestor from 6 to 7 million years ago, and their DNA is 98–99% like ours. This close connection helps explain why their behavior, tool use, problem-solving skills, emotional displays, cooperation, and conflicts seem so familiar. Key differences, such as the absence of symbolic language or cumulative culture, highlight what is uniquely human.

Chimpanzees possess a powerful build, featuring long arms, grasping hands and feet, and expressive faces. Adults typically weigh 70 to 130 pounds, with males being larger and more robust than females. In the wild, they live 30 to 40 years, and considerably longer in protected settings.

Chimpanzees inhabit a wide range of environments, from dense tropical forests to open woodlands, adapting their diets and behaviors to local conditions. Today, habitat loss, hunting, and disease threaten their survival, making conservation efforts, many inspired or led by Jane, critical to their future.

---

**SCIENCE SPOTLIGHT**

## What Chimpanzees Reveal About Human Evolution

Chimpanzees are our closest living relatives, and because we share a common ancestor from 6 to 7 million years ago and 98–99% of our DNA, their lives offer a rare window into the deep evolutionary roots of human behavior.

Many chimpanzee traits, alliances, rivalries, territorial defense, cooperation, and reconciliation show striking parallels with human social strategies. Other features, such as the lack of symbolic language or large-scale cultural institutions, reveal where the human lineage diverged. These similarities and differences help scientists identify which aspects of behavior are ancient and which evolved later.

Long-term observations from Gombe, including teaching, cultural traditions, maternal learning, and occasional violence, have become essential for understanding the evolution of aggression, empathy, social bonding, problem-solving, and community structure. By comparing chimpanzee behavior with our own, researchers can trace how human social life emerged from our primate past.

Chimpanzees remain invaluable to evolutionary science because they illuminate both the continuities we share and the distinctive pathways that shaped the human mind.

## David Greybeard and a Scientific Breakthrough

In late 1960, Jane discovered that would change everything. While observing David Greybeard, she witnessed behavior never before documented in wild chimpanzees. On November 4, 1960, she watched him use a grass stem to extract termites from a mound. Shortly after, he grabbed a twig, removed the leaves, and tweaked it to work better.

This was the first recorded scientific observation of a wild chimpanzee modifying and using a tool. At the time, toolmaking was considered a defining human trait. Jane waited until she had observed the behavior repeatedly before reporting it. When she finally informed Louis Leakey, his response was immediate and famous: "Now we must redefine man, redefine tool, or accept chimpanzees as humans."

The discovery transformed scientific understanding. It challenged the belief that humans alone possessed the ability to create and use tools and secured long-term support for Jane's research.

Her observations did not stop there. She documented chimpanzees hunting, sharing food, and cooperating within their groups, behaviors that further blurred the line between human and animal. Later, Jane observed chimpanzees using leaf sponges to collect water, stone hammers to crack nuts (in other populations), and even spear-like sticks to hunt bushbabies (discovered later by other researchers). She also found that different chimpanzee communities used different tools, suggesting that chimpanzees possess culture: behaviors learned socially rather than inherited genetically.

**SCIENCE SPOTLIGHT**

## David Greybeard: The Chimpanzee Who Changed Science

David Greybeard was the first wild chimpanzee Jane Goodall observed using a tool, an act that overturned one of science's most deeply held assumptions. At that time, toolmaking was considered a defining human trait. When Jane watched David strip leaves from a twig and use it to "fish" termites from their mound, he showed that another species could shape and use objects with intention. Louis Leakey's response became legendary: "Now we must redefine man, redefine tool, or accept chimpanzees as humans."

Termite fishing itself is a simple but ingenious sequence. A chimpanzee selects a slender twig, removes its leaves, and inserts it into the narrow tunnels of a termite mound. Soldier termites clamp onto the intruding stick; when the chimp withdraws it, they are lifted out and eaten. The behavior requires choice, modification, and technique, clear evidence of problem-solving, and cultural learning.

David Greybeard's importance extended far beyond this single discovery. In the early Gombe years, he was the first chimpanzee to accept Jane's presence, approaching her calmly when others fled. His trust opened the door to the entire community. Because he tolerated her, other chimpanzees gradually followed his lead, allowing Jane to observe their daily lives, relationships, and social structure. Without David's quiet confidence, the long-term study that transformed primatology might never have taken root.

David Greybeard's groundbreaking tool use and his early bond with Jane reshaped scientific understanding of animal minds and helped reveal the deep evolutionary connections between humans and our closest living relatives

**SCIENCE SPOTLIGHT**

### Termite Fishing and Chimpanzee Tool Culture

Termite fishing is one of the most distinctive examples of chimpanzee ingenuity. In this behavior, a chimpanzee selects plant material, prepares it, and inserts it into the tunnels of a termite mound to draw out soldier termites. The technique varies between communities: some prefer grass stems, others modify twigs, showing that the skill is learned, practiced, and refined over time.

What makes termite fishing so significant is not just the act itself but the variations surrounding it. Young chimpanzees learn the technique by watching their mothers, and different groups use different tools or methods, revealing patterns of cultural transmission. These community-specific traditions extend beyond termite fishing to include leaf sponging for water, nut cracking with stones, and other specialized foraging strategies.

Together, these behaviors show that chimpanzees have flexible problem-solving abilities and pass knowledge across generations, providing evidence of a rich tool culture once thought to be uniquely human.

## Individuals, Not Just Animals

As the chimpanzees grew accustomed to her presence, Jane recognized them as individuals. Unlike most scientists of her time, she gave them names rather than numbers.

David Greybeard was the first chimpanzee to accept Jane's presence. Gentle and patient, he became the individual who revealed tool use to the world. Mike, a smaller but highly intelligent male, later astonished researchers by using empty kerosene cans to intimidate rivals and seize the alpha position from Goliath, a larger and previously dominant male. Each chimpanzee displayed a distinct personality, convincing Jane that they were individuals rather than interchangeable members of a group. One of the most important chimpanzees that Jane met was Old Flo.

Old Flo was a high-ranking, deeply maternal female who became the matriarch of a well-studied family line. Recognizable by her bulbous nose and ragged ears, Jane observed her interactions with her children, including Fifi and Figan, documenting relationships that revealed deep maternal bonds and long-term social influence. She was the mother of Fifi (born 1958), Faben (born 1960), Figan (born 1964), Freud (born 1971), and Flint (born 1975), all of whom became central figures in Gombe's long-term story.

Jane followed this family for years, gaining rare insight into chimpanzee life across generations. Chimpanzees typically live 30 to 40 years in the wild but can survive into their 50s or 60s in protected settings.

When Old Flo died in 1972, the impact was profound. Her young son Flint, still dependent on her, died shortly afterward, an apparent case of grief so severe that he could not survive without her.

The event resonated beyond the scientific community. Old Flo's fame grew so extensive that The Times of London reported her death.

Several of Old Flo's descendants would later become dominant males within the group, and researchers have since followed many of Old Flo's descendants across multiple generations, creating one of the longest continuous family records ever assembled for a wild animal population. Among Old Flo's descendants was Frodo, Fifi's powerful and aggressive son, who became one of Gombe's most famous chimpanzees. Notorious for attacking rival chimpanzees and occasionally human observers, including Jane herself, Frodo showed both the strength and unpredictability of wild chimpanzee behavior.

Together, these chimpanzees formed the first generation of individuals through whom Jane revealed the emotional depth, intelligence, and social complexity of wild chimpanzee life.

Figure 2. Two young chimpanzees in Gombe National Park, Tanzania. Photograph by Cethuyghe. Licensed under CC BY-SA 4.0. Adapted for publication in black and white.

**SCIENCE SPOTLIGHT**

## Chimpanzee Society

Chimpanzees live in large, fluid communities, often of 30 to over 100 individuals, organized around shifting alliances, family bonds, and long-term relationships.

The central figure in each community is a dominant male, whose status is determined by strength, political acumen, the ability to form alliances, and the backing of crucial supporters. Females form their own networks, often centered on maternal lines, and play crucial roles in stability and knowledge transmission. Young chimpanzees learn by watching their mothers and older siblings, acquiring skills from tool use to social etiquette over many years.

Territory is essential. Communities defend defined home ranges, patrolling borders and sometimes engaging in coordinated aggression against neighboring groups. Within their own community, however, chimpanzees maintain relationships through grooming, play, reconciliation, and shared activities that reinforce trust.

Chimpanzee societies are dynamic, strategic, and highly complex social communities that reveal how cooperation, conflict, and culture evolved long before humans appeared. Their ability to form alliances, transmit knowledge, balance cooperation with conflict, and pass behaviors from one generation to the next provides valuable insights into the evolutionary origins of human society.

## Married Life

By the early 1960s, Jane's discoveries, tool use, termite fishing, and meat-eating among chimpanzees, were reshaping scientific understanding. Recognizing the significance of her work, Leakey approached National Geographic, the primary funder of her research, and requested visual documentation of her methods and findings.

The magazine agreed and selected Hugo van Lawick, a 25-year-old Dutch wildlife photographer known for his patience, technical skill, and resilience in remote environments. Leakey formally invited him to join the Gombe project, and in June 1962, Hugo arrived in Tanzania.

Jane was initially hesitant. She feared that the presence of another person might disrupt the fragile trust she had built with the chimpanzees. But she soon recognized Hugo's deep respect for animals and his ability to move quietly within the environment. The two quickly formed a strong connection.

They shared a love of nature and life in the field, and their partnership grew into a relationship. On March 28, 1964, they were married at the Chelsea Registry Office in London. Jane became Baroness Jane van Lawick-Goodall.

Through his photography and films, Hugo played a crucial role in bringing Jane's work to a global audience. His images helped people see chimpanzees and Jane's work in a new and compelling way.

Yet the very work that brought them together also pulled them apart. Jane remained deeply committed to her research at Gombe, while Hugo's career required travel to other locations, including the Serengeti. As their professional paths diverged, so did their daily lives. "I built up the research station and I had to stay," Jane later reflected. "That was my work."

Over time, distance and competing demands strained the marriage. They began to spend long periods apart, and the relationship gradually weakened. In 1974, after ten years of marriage, they divorced.

Despite the separation, they remained on good terms. Jane later described the marriage as meaningful, even though they could not sustain it under the pressures of two demanding and divergent careers.

## First Cover Story and Film

In 1963, *National Geographic* published its first cover story on Jane and her research, titled *My Life Among the Wild Chimpanzees.* It marked the first major global introduction to her work at Gombe and brought her observations into homes around the world.

The article featured photographs by Hugo, which quickly became iconic. For many readers, it was the first time they saw chimpanzees presented not as distant animals, but as individuals with personalities, relationships, and emotional lives.

The impact was immediate. Public interest surged, transforming what had been a quiet, remote field of study into an internationally recognized scientific breakthrough. The exposure also helped secure long-term funding for the Gombe project.

This moment marked a turning point. Jane was no longer just a field researcher; she had become a communicator, a bridge between science and the public, and an emerging global figure.

The National Geographic documentary *Miss Goodall and the Wild Chimpanzees*, first broadcast on American television on December 22, 1965, introduced the wider public to Jane's work.

Filmed by Hugo and narrated by Orson Welles, the documentary brought viewers into the forests of Gombe. It showed a quiet but determined junior researcher patiently observing chimpanzees in their natural habitat, often for hours at a time, waiting for brief but meaningful moments of behavior.

For many, this was their first actual glimpse into the lives of wild chimpanzees. The film presented them not as distant or primitive creatures, but as social, expressive beings with recognizable behaviors and relationships.

The impact was significant. The chimpanzees of Gombe soon became familiar to audiences in both the U.S. and Great Britain, appearing regularly in documentaries and television programs. Through these films, Jane's work reached far beyond the scientific community.

Hugo's role was critical. His photographs and footage provided visual evidence that supported Jane's observations, evidence that many scientists had initially questioned. As Jane later reflected, it was through his images that others accepted what she had been describing: that chimpanzees share gestures, expressions, and social bonds strikingly like our own.

Together, their work helped shift public and scientific perception, challenging long-held assumptions about the divide between humans and other primates.

---

**SCIENCE SPOTLIGHT**

## How Chimpanzees Communicate

Chimpanzees communicate through a rich blend of sounds, gestures, facial expressions, and body language. Their most iconic call is the pant-hoot, a rising, rhythmic vocalization used to signal excitement, rally group members, or reinforce bonds across long distances. In tense moments, they produce sharp screams, while softer grunts convey reassurance or mild interest during peaceful interactions.

Gestures are equally important. A lowered posture, a gentle touch, or an extended hand invites grooming, a behavior crucial for building trust within chimpanzee society. Chimpanzees also use embraces to comfort others or reaffirm alliances. Hand-clasp grooming is one of their most distinctive gestures, and two chimpanzees perform it by raising an arm and clasping hands while grooming with the other. Communities vary in this act, and they consider it a cultural tradition.

Facial expressions add nuance: a relaxed, open-mouth grin signals playfulness, while a tense, bared-teeth face shows fear or submission. Body language also carries meaning; males perform dramatic dominance displays, charging, stamping, and dragging branches, to assert status or intimidate rivals.

## Studying for a PhD

As Jane's discoveries gained attention, it became increasingly clear that formal academic recognition would strengthen their scientific impact. Leakey arranged funding for her to pursue a doctorate at Newnham College, University of Cambridge, England.

Her admission in 1962 was highly unusual. Jane did not hold an undergraduate degree, yet she became one of the very few, reportedly the eighth person, allowed to enter Cambridge directly as a PhD candidate.

During her studies, Jane divided her time between Cambridge and Gombe. She spent extended periods in the field collecting data, then returned to Cambridge for supervision, seminars, and writing, before heading back to Africa to continue her observations.

In her absence, her local field assistants played an increasingly important role. They maintained daily observations, recorded chimpanzee activity, and ensured continuity in the long-term study. Over time, they developed deep expertise in tracking chimpanzees and interpreting behavior in the forest. Their contributions became an essential part of the data supporting her research.

At Cambridge, Jane faced criticism from more traditional scientists who challenged her for naming the chimpanzees rather than assigning them numbers, and for describing their behavior using terms such as "childhood," "emotion," and "motivation." Traditionally, scientists numbered animals to avoid attributing human emotions or personalities. Jane argued that recognizing each chimpanzee as an individual was necessary to understand their unique traits and social relationships. Many now widely accept her approach in long-term behavioral studies. Jane did not change her approach. She believed that what she observed justified those descriptions, that the chimpanzees were not simply reacting but experiencing.

In 1965, she completed her thesis, *Behaviour of Free-Living Chimpanzees*, under the supervision of Robert Hinde. Based on her first five years at Gombe, she documented the complex social lives, behaviors, and relationships she had uncovered.

She received a PhD in ethology the same year, formally establishing herself as a scientist, though her work had already begun reshaping the field long before.

## Grub's Upbringing

On March 4, 1967, Jane gave birth to her son, Hugo Eric Louis van Lawick, in Nairobi, Kenya. Jane affectionately nicknamed him "Grub," and the nickname stayed with him throughout his life.

Grub spent part of his early childhood at Gombe, growing up alongside his mother's research. Rather than step away from her work, Jane brought him with her into the field, a decision that was both unconventional and demanding.

The environment, however, was not without risk. Chimpanzees, though familiar, were still wild animals capable of aggression. To ensure his safety, Jane never left Grub unattended and constructed a large protective enclosure for him when he was an infant. It resembled an oversized cot, spacious enough for movement while keeping him secure.

Motherhood deepened Jane's scientific observations. Her close study of chimpanzee families shaped how she understood and approached raising her own child. She had already documented differences between experienced chimpanzee mothers, who calmly introduced their infants to the group, and first-time mothers, who often hid their babies, triggering agitation within the troop.

She also observed the strength of the maternal bond and mirrored this in her own life. Young chimpanzees remain dependent on their mothers for five to eight years, one of the longest childhoods among mammals, giving them time to learn complex social and survival skills. As a result, for the first three years, she rarely separated from her son for more than a few hours.

Grub's early experiences reflected his unusual upbringing. Surrounded by wildlife, his first sentence, "That big lion out there eat me," captured both the environment he lived in and the awareness it demanded.

As he grew older, his family sent Grub to England for schooling, and he spent much of the year in Bournemouth with his grandmother, returning to Tanzania during school holidays to be with his mother.

In adulthood, he chose a different path. Grub settled in Tanzania, where he and his wife, Maria, raised their three children: Angel, Merlin, and Nick. He became a boat builder, remaining close to the natural world, though not pursuing scientific research.

Jane later reflected that while her son did not follow her career, his early years gave him something equally valuable: a deep familiarity with nature and a lasting appreciation for the freedom and beauty of that life.

Balancing motherhood with fieldwork, and later, an increasingly demanding global schedule, was never simple. Yet Jane resisted the expectation that women had to justify such choices in ways men did not have to, continuing to shape both her family life and her scientific legacy on her own terms.

---

**SCIENCE SPOTLIGHT**

## Do Chimpanzees Have Culture?

Chimpanzees live in communities that develop their own ways of doing things, patterns of behavior that persist over time and differ from group to group. These are best understood as traditions and habits that become part of a community's identity. Some groups favor certain grooming styles, others use distinctive foraging techniques, and still others develop unique social rituals that neighboring communities do not share.

Many of these customs begin early in life. Young chimpanzees spend years closely observing their mothers, absorbing the subtleties of daily life: how to handle food, how to approach others, how to navigate conflict, and how to take part in shared activities. Since they learn these behaviors rather than inheriting them, they can vary widely across regions.

Researchers have documented striking regional differences in everything from tool preferences to greeting gestures. One community may use leaves in a particular way, while another develops a different method entirely. These variations arise through social learning, as individuals watch, imitate, and practice the behaviors of trusted group members.

These shared practices, when considered as a whole, create a rich fabric of community-specific customs. They show chimpanzees do not simply react to their environment; they build traditions, pass them on, and shape the social worlds they inherit.

## Gombe Stream Research Center

When Jane first arrived, the British colonial administration had designated Gombe as the Gombe Stream Game Reserve, a protected area established in 1943. While intended to safeguard wildlife, it still permitted limited, regulated hunting.

In 1967, following the completion of her PhD, Jane returned to Gombe to continue her research and helped establish the Gombe Stream Research Center (GSRC). The center coordinated the growing body of chimpanzee research and, to support the increasing number of scientists, many of them women, drawn to study primates, ecology, and field-based behavioral science using her methods.

In 1968, the year after, the Tanzanian government elevated the reserve to a national park and renamed it Gombe Stream National Park. The change reflected both stricter conservation protections and the global importance of Jane's work. As part of a broader standardization in 1973, they simplified the name to Gombe National Park. Despite its scientific importance, Gombe remains one of the smallest national parks in Tanzania, covering just 35 square kilometers along the eastern shore of Lake Tanganyika.

Over time, the research center evolved into a permanent institution. Today, trained Tanzanian researchers and field assistants largely operated it, reflecting a shift toward local leadership in long-term scientific work. The Gombe study is now the longest-running continuous field study of any animal species in its natural environment, spanning over 60 years. This extraordinary dataset has provided rare insights into chimpanzee life across generations, including social hierarchies, maternal behavior, hunting patterns, and what some researchers describe as cultural transmission.

The center has also expanded its focus to include broader ecological and conservation research, studying threats such as disease, habitat loss, and human encroachment. It has become a training ground for generations of primatologists and conservation biologists.

## The Banana Controversy

As Jane's work gained prominence, it also attracted criticism. In the early 1970s, researchers questioned the use of provisioning, the practice of placing bananas at specific locations to attract chimpanzees and establish consistent observation points.

Critics argued that provisioning altered natural behavior, increasing aggression and disrupting social dynamics. Jane acknowledged these concerns but maintained that in the early stages of the study, feeding had been necessary to habituate the chimpanzees to human presence and to allow close observation.

Over time, however, her approach evolved. By the mid-1970s, the use of feeding stations had decreased, and by the late 1970s, they had largely eliminated them. The research shifted toward fully non-provisioned observation, prioritizing natural behavior over accessibility.

The debate did not disappear entirely. Some researchers have suggested that early human intervention may have intensified certain behaviors observed at Gombe, particularly heightened aggression and intergroup conflict. Others point out that similar patterns have since occurred in chimpanzee populations where no provisioning happened. Jane herself took a balanced view. She accepted that feeding may have influenced the intensity of certain behaviors but argued that it did not fundamentally alter the nature of chimpanzee social life.

Another concern that emerged during this period was disease transmission. As understanding grew, researchers recognized that close contact and shared feeding sites could increase the risk of spreading human illnesses to chimpanzees. This led to the introduction of stricter health protocols, further reinforcing the move away from provisioning.

Today, the controversy is viewed as part of the natural evolution of field science, an early methodological compromise that was later refined as knowledge and standards advanced.

## The Darker Side of Chimpanzee Life

During her first decade at Gombe, Jane came to believe that chimpanzees were, in many ways, gentler than humans. She had observed affection, hugs, grooming, playful interactions, and strong social bonds within the troop. But over time, a more complex picture emerged.

Alongside these cooperative behaviors, Jane documented clear evidence of aggression and violence. Chimpanzees hunted and killed other animals, particularly red colobus monkeys, often working together in coordinated groups. She observed hunts in which individuals blocked escape routes while others climbed to capture prey, followed by the sharing of meat within the group.

She also witnessed troubling behavior within the troop itself. Dominant females sometimes killed the infants of other females and occasionally ate them. These incidents did not occur in isolation but formed a broader pattern, which challenged earlier assumptions.

The most disturbing events at Gombe began in 1974, when a once unified chimpanzee community fractured into two groups: the Kasakela in the north and the Kahama in the south. What followed was what Jane later called the "Four-Year War." Over several years, males from the larger Kasakela group formed coordinated coalitions and carried out a series of deliberate attacks on isolated Kahama males. The Kasakela group eventually killed or drove off the Kahama chimpanzees one by one and then expanded their territory into the area the Kahama had once occupied.

For Jane, the experience was profoundly unsettling. The behavior she observed, which included planned ambushes, group aggression, and territorial takeover, was like the organized, coalition-based violence seen in human warfare. "I was shocked. I was saddened," she later said. "But I realized… this makes them even more like us than I thought before."

The conflict eliminated the Kahama community and allowed the Kasakela chimpanzees to occupy their former territory, providing one of the first documented examples of coalition-based intergroup warfare in a non-human species.

Scientifically, the Four-Year War marked a major shift in understanding. Chimpanzees were not simply peaceful or violent; they were capable of both remarkable cooperation and organized, lethal conflict. These observations opened new avenues of research into the evolutionary origins of aggression, alliance formation, territorial behavior, and warfare, suggesting that some aspects of human conflict may have deep roots in our shared primate ancestry.

## Derek Bryceson

In 1975, one year after her divorce from Hugo, Jane married Derek Bryceson. A former Royal Air Force pilot during World War II, Derek was also a prominent figure in Tanzania's government, serving as Director of the Tanzania National Parks Authority (TANAPA) and as a member of parliament. Through his role, Derek had direct influence over conservation policy, park management, and anti-poaching efforts. He became an important ally in supporting and protecting Jane's work at Gombe.

Their life together in Tanzania was relatively quiet. They shared a deep commitment to conservation and often traveled together, sometimes flying over national parks in Derek's small plane. Just weeks after their marriage, however, events took a dramatic turn.

On May 19, 1975, armed rebels crossed into Tanzania from Zaire (now the Democratic Republic of the Congo) and kidnapped four researchers from Gombe: three American students and one Dutch researcher. The attackers forced them across Lake Tanganyika.

Jane was at the site during the attack and escaped with other staff members. In the aftermath, she worked closely with Tanzanian authorities to coordinate the response, while Derek used his political position to help secure the hostages' release. After 13 days, they freed the captives.

The incident had lasting consequences. Researchers at Gombe temporarily suspended their work, people evacuated the camp, and many international researchers left the area. The park authorities strengthened security measures, imposing tighter controls on access and movement within the park.

Despite these challenges, Jane continued her work, increasingly relying on trained local field assistants to maintain the continuity of the research. Derek remained a steady source of support during this period. His influence helped ensure the long-term protection of Gombe and the stability of Jane's research efforts.

In October 1980, after five years of marriage, Derek died following a prolonged illness. His death was a profound loss. Afterward, Jane divided her time between Tanzania and England, continuing the work that had come to define her life.

## The Jane Goodall Institute

By the mid-1970s, Jane saw a troubling transformation around Gombe. Despite unchanging park boundaries, the forest outside them vanished quickly as people cleared trees for farmland and firewood. Gombe was becoming an island of green surrounded by bare, eroded hills. As habitats vanished, chimpanzee territories shrank, food sources declined, and populations became more vulnerable to disease and inbreeding. Jane later described flying over the region and seeing dense forest inside the park and barren land beyond it.

Her growing international travel exposed her to wider threats: poaching, the bushmeat trade, and the capture of orphaned chimpanzees. These experiences marked a turning point, shifting her focus from observation to protection.

In 1977, she founded the Jane Goodall Institute to support long-term research at Gombe, protect wildlife habitats, and work with local communities on sustainable conservation. Rather than imposing outside solutions, Jane emphasized listening to local needs and building programs that communities would adopt. Forest regeneration began, and people became active partners in conservation.

The Institute expanded beyond Gombe, supporting sanctuaries, anti-poaching efforts, conservation programs, and training for local leaders in Africa. Its early headquarters in the San Francisco Bay Area reflected Jane's increasing presence in the United States for fundraising, scientific collaboration, and public advocacy. The Institute also became steward of the Gombe Stream Research Center, ensuring the continuation of the world's longest-running chimpanzee study.

Following its creation, Jane's role evolved once again, from field researcher to global advocate.

## A Turning Point

By 1980, the Institute's growth required professional management. Jane was traveling constantly, speaking, writing, and raising awareness about habitat loss and wildlife protection. Recognizing where she could have the greatest impact, she stepped back from day-to-day operations. A dedicated team took over administration and long-term planning, allowing the institute to expand sustainably. Jane remained its founder and guiding voice, focusing on advocacy, education, and global outreach. This ensured that research at Gombe continued uninterrupted and that conservation programs could grow worldwide.

In 1986, after 25 years of research, she published *The Chimpanzees of Gombe: Patterns of Behavior*, a landmark scientific work. That same year, at the "Understanding Chimpanzees" conference in Chicago, she confronted the accelerating destruction of chimpanzee habitats, the threat of extinction, and the suffering of chimpanzees in laboratory research. She later said, "I went to the conference as a scientist... and I left as an activist." From that moment, her focus shifted decisively to global conservation.

She stepped away from full-time fieldwork and committed herself to raising awareness, advocating for environmental protection, and campaigning against the exploitation of chimpanzees. She came to understand that saving chimpanzees required addressing the broader relationship between humans and the natural world.

---

**HISTORICAL SPOTLIGHT**

### Gary Larson cartoon incident

In 1987, Gary Larson published a Far Side cartoon showing two chimpanzees grooming, with one discovering a blonde hair and saying, "Conducting a little more 'research' with that Jane Goodall tramp?" Unaware that Jane was in Africa, the Jane Goodall Institute denounced the cartoon as an "atrocity" in a lawyer-drafted letter to Larson and his syndicate.

When Jane later saw it, she found it funny, and it eventually became her favorite pop culture portrayal of herself. Larson donated the profits from a T-shirt featuring the cartoon to the Institute, and Jane later wrote the preface to The Far Side Gallery 5, recounting the misunderstanding and praising Larson's imaginative comparisons between humans and animals. When Larson visited Gombe in 1988, Frodo, one of the more aggressive chimpanzees, promptly charged him.

## Speaking Out for Chimpanzee Welfare and the Environment

By the late 1980s, people widely regarded Jane as the world's leading authority on chimpanzees. In 1989, she appeared at a press conference organized by the World Wildlife Fund in Washington, D.C., where the United States announced its decision to list all African chimpanzees as endangered under the Endangered Species Act.

From this point onward, Jane's work increasingly focused on advocacy. She spoke out against the use of chimpanzees in scientific research, campaigning for improved conditions and stricter limits on their use. She also traveled extensively across Africa, visiting rescue centers and working with governments to promote conservation and nature-based tourism.

In 1992, she founded the Tchimpounga Chimpanzee Rehabilitation Center in the Republic of Congo. The sanctuary cared for chimpanzees orphaned by the bushmeat trade and now houses more than a hundred animals across several protected islands.

Jane's advocacy extended beyond individual sanctuaries. She drew global attention to the destruction of chimpanzee habitats through deforestation, poaching, and human expansion. Much of this habitat lies outside protected national parks, making local community involvement essential. Working with scientists and organizations, including collaboration with NASA, she supported the use of satellite imagery to help communities manage land use and reduce environmental damage.

By the early 2000s, Jane was devoting nearly all her time to conservation, traveling close to 300 days a year to speak, educate, and advocate for wildlife and the environment.

## Supporting the Ethical Treatment of Animals

Jane became one of the most influential voices calling for the humane treatment of animals, particularly those used in research. After visiting laboratory facilities in the 1980s, she pushed for introducing enrichment programs, activities designed to improve the psychological well-being of captive chimpanzees. Her involvement helped shift standards within research institutions.

She argued scientists must work toward alternatives to animal testing, while also emphasizing the importance of constructive dialogue rather than confrontation. Jane openly opposed extremist tactics within the animal rights movement, believing they undermined progress and polarized debate.

Her advocacy later expanded to include zoo animals, farm animals, and broader issues of animal welfare. In 1986, she helped establish the Committee for the Conservation and Care of Chimpanzees, and she continued to support sanctuaries and rehabilitation programs worldwide.

Jane also used writing as a tool for education, including children's books such as *The Chimpanzee Family Book*, helping to shape public understanding of animals as intelligent, emotional beings deserving of respect.

A lifelong advocate for ethical living, Jane adopted a vegetarian and later vegan lifestyle, linking diet to environmental sustainability, animal welfare, and human health. She consistently encouraged individuals to consider the broader impact of their choices, reinforcing her central message: that protecting animals and the environment begins with human responsibility.

## TACARE – Community-Led Conservation

In the early 1990s, another flight over Gombe revealed the full extent of deforestation beyond the park's borders. Gombe's boundaries protected it, but the outside increasingly made it vulnerable. Jane realized that conservation would fail if people in surrounding communities did not have their needs met.

She began by listening. Working with Tanzanian villages, she helped identify the challenges they faced: poverty, limited resources, and lack of education, and explored solutions that supported both people and the environment. Gradually, forests recovered.

This approach became the foundation of TACARE (Lake Tanganyika Catchment Reforestation and Education), developed in the 1990s with Emmanuel Mtiti. TACARE's foundation rested on a simple principle: local communities must benefit so that conservation can succeed. Instead of top-down directives, it emphasized partnership, supporting sustainable agriculture, reforestation, education, and healthcare. The results were gradual but enduring. Forest cover expanded, and the program grew to over 100 villages, influencing community-led conservation efforts across Africa.

Alongside environmental work, Jane championed education for girls in Tanzania and Uganda, supporting scholarships, health resources, and peer networks. She understood that empowering communities, especially women and young people, was essential to protecting the natural world.

## Roots & Shoots – Inspiring the Next Generation

In 1991, Jane turned her attention to the next generation, launching the Roots & Shoots program through the Jane Goodall Institute. The program began modestly, with a group of twelve students meeting at her home in Dar es Salaam. They spoke about the challenges they saw in their communities: environmental destruction, poverty, and social issues, and they wanted to act.

The founders established Roots & Shoots based on a simple idea: meaningful change begins locally. The program encouraged young people to identify problems in their own communities and take practical steps to address them, whether through environmental projects, wildlife protection, or humanitarian work. Quickly expanding, the program fostered leadership, critical thinking, and a sense of responsibility. It required no special resources or permission, only the willingness to act.

Over time, Roots & Shoots grew into a global movement, reaching over 75 countries, with thousands of active groups and over a million participants. In 2004, Kofi Annan, who honored her contributions to building a more peaceful and sustainable world, recognized Jane's work with the program.

## Spreading the Word

By the later years of her life, Jane had become not only a scientist but a global advocate. She spent much of each year traveling, often close to 300 days, speaking to audiences around the world about conservation, animal welfare, and the future of the planet. Between journeys, she returned to her childhood home in England, maintaining a quiet base amid an otherwise demanding schedule.

Despite the constant travel, her work was deeply personal. When asked whether she preferred chimpanzees or people, Jane often responded with characteristic honesty: "Chimpanzees are so like us… that I like some people much more than some chimpanzees, and some chimpanzees much more than some people."

Her message drew on decades of observation, but also on the influence of earlier environmental thinkers such as Rachel Carson. While attending the University of Cambridge in the 1960s, Jane read Silent Spring and the book's warning about environmental damage caused by pesticides deeply affected her. Carson's willingness to challenge governments and industry left a lasting impression.

Over time, Jane expanded her message beyond chimpanzees to the broader relationship between humans, animals, and the environment. She spoke about the interconnectedness of ecosystems and the consequences of human actions, from deforestation and the bushmeat trade to industrial farming and pollution.

She also became an advocate for more ethical lifestyles, including reducing meat consumption, linking personal choices to environmental sustainability and animal welfare. Central to her message was a belief in individual responsibility: that meaningful change begins with everyday actions.

She remained realistic about the challenges. Chimpanzee populations continued to decline because of habitat loss, hunting, and the illegal pet trade. Yet Jane consistently emphasized hope, pointing to the capacity of individuals, especially young people, to make a difference through programs like Roots & Shoots.

She often shared stories that illustrated this belief: ordinary people taking small but meaningful actions that created real change. These examples reinforced her core message that compassion, when put into practice, has the power to reshape the world.

## Honors and Recognition

Throughout her later life, Jane received many awards and honors in recognition of her contributions to science, conservation, and global advocacy. These ranged from international prizes to academic distinctions, reflecting the breadth of her impact.

She received 17 honorary degrees from universities around the world, besides major conservation and scientific honors, including the Gold Medal of Conservation from the San Diego Zoological Society (1974), the J. Paul Getty Wildlife Conservation Prize (1984), the Schweitzer Medal from the Animal Welfare Institute (1987), the Centennial Award from the National Geographic Society (1988), and the Kyoto Prize in Basic Sciences (1990).

In 1995, the United Kingdom appointed Jane Commander of the Order of the British Empire (CBE) for her services to zoology. They later elevated her to Dame Commander of the Order of the British Empire (DBE) in 2003. In 2004, she received a formal investiture at Buckingham Palace for her scientific achievements and global conservation leadership.

Her international recognition continued to grow. The United Nations named her a Messenger of Peace in 2002, recognizing her work in environmental protection and human welfare. In 2006, France's highest order of merit, the Légion d'Honneur, appointed her an Officer for her pioneering research and dedication to conservation.

Her influence has gained wide acknowledgment in more recent years. *Time* named her one of the 100 most influential people in the world in 2019. In 2022, she received the Stephen Hawking Medal for Science Communication. In 2025, the United States awarded her the Presidential Medal of Freedom, one of the highest civilian honors.

These recognitions reflect not only her scientific achievements but also her lasting role as a global voice for environmental stewardship.

---

**SCIENCE SPOTLIGHT**

## What Is Prosopagnosia?

Prosopagnosia, often called face blindness, is a neurological condition that makes it difficult to recognize faces, even those of familiar people. Individuals with prosopagnosia usually have normal vision and intelligence but often rely on other clues, such as a person's voice, hairstyle, posture, or mannerisms, to identify them. The condition may be present from birth or develop after injury to areas of the brain involved in facial recognition.

Later in life, Jane Goodall realized she experienced prosopagnosia after meeting others with similar experiences, and neurologist Oliver Sacks subsequently diagnosed the condition. Her younger sister also shared it, suggesting a possible inherited component.

Despite sometimes struggling to recognize human faces, Jane had little difficulty distinguishing individual chimpanzees. Years of careful observation enabled her to identify subtle differences in their facial features, expressions, movements, and behavior. This remarkable contrast highlights both the adaptability of the human brain and Jane's extraordinary observational skills, qualities that played a key role in her groundbreaking research.

---

## Writing and Public Works

Alongside her scientific research, Jane was a prolific writer and communicator. Her first major book, In the Shadow of Man (1971), introduced her work at Gombe to a global audience, and translators converted it into dozens of languages later.

She followed this with *The Chimpanzees of Gombe: Patterns of Behavior* (1986), a comprehensive scientific account of her early research, as well as more personal and reflective works such as *Reason for Hope: A Spiritual Journey* (2000) and *The Book of Hope: A Survival Guide for Trying Times* (2021).

Over the course of her career, Jane authored over 30 books, including many written for children, helping to educate and inspire younger generations. Her writing consistently emphasized a central theme: that protecting the natural world requires both understanding and compassion.

## Media and Public Recognition

Public awareness of Jane's work began in 1965 with the National Geographic film *Miss Goodall and the Wild Chimpanzees* and the accompanying magazine coverage. These early productions introduced global audiences to the chimpanzees of Gombe and established Jane as a compelling new voice in science.

In 1988, Hugo's documentary *People of the Forest* brought British viewers into the lives of Gombe's "F" family: Flo, Fifi, Flint, and their descendants. Released internationally in 1991, the film deepened the public's understanding of chimpanzee social life and showcased the richness of long-term field research. By the time Hugo stopped filming at Gombe, he had created a visual archive spanning over twenty years, capturing relationships, births, deaths, and generational change within one of the world's most closely studied chimpanzee communities.

Jane's story reached a new generation in 2017 with the National Geographic documentary *Jane*. Built from previously unseen archival footage, the film offered an intimate portrait of her early years in Gombe and traced her evolution into a leading conservationist. Widely acclaimed and the recipient of multiple awards, it renewed global interest in her life and work. Jane actively took part in promoting the film, often appearing with her son, Grub, and connecting with audiences of all ages.

Her work has continued to be documented through many films and television specials, including *Jane's Journey*, *Jane Goodall's Wild Chimpanzees*, and the 2023 documentary *Reasons for Hope*, which highlights her four pillars of hope: human intellect, the resilience of nature, the energy of young people, and the indomitable human spirit. She has also reached younger viewers through educational programming, including the Apple TV+ series *Jane*, where she appears as herself.

Filmmaking over the decades has brought Jane's research and message to a global audience, ensuring the Gombe chimpanzees and their lessons remain part of our shared cultural imagination.

## Death and Legacy

Jane remained active in her work well into her later years, continuing to travel extensively, speak publicly, and advocate for conservation around the world. She spent much of her time encouraging governments, organizations, and individuals to protect wildlife and preserve natural habitats.

She died on October 1, 2025, at 91 years old, in Los Angeles, California, from natural causes. At the time of her death, she was still engaged in her work, scheduled to meet with students and educators as part of a conservation initiative.

Jane's legacy extends far beyond her early research at Gombe. Through her groundbreaking observations, she transformed the scientific understanding of chimpanzees, demonstrating that they are intelligent, emotional beings with complex social lives. Her work helped reshape how scientists and the public view the relationship between humans and other animals.

Through the Jane Goodall Institute, she built a global network supporting wildlife research, habitat protection, and community-based conservation. Her approach highlighted the inseparability of animal protection and support for local communities.

Over time, her influence expanded from science to global advocacy. She became a leading voice for environmental protection, animal welfare, and sustainable living, inspiring generations to take an active role in protecting the natural world. Jane often encouraged young people to believe that individual actions matter. She emphasized that meaningful change does not require grand gestures but begins with small, consistent efforts.

Her work continues today through ongoing research at Gombe, global conservation programs, and initiatives such as Roots & Shoots, which empower individuals to take responsibility for their local environments.

Jane's life showed that curiosity, patience, and persistence can change how the world understands itself.

---

**HISTORICAL SPOTLIGHT**

## Jane Goodall's Final Live Public Appearance (2025)

A reflective farewell delivered at the Bloomberg Philanthropies Global Forum

In September 2025, during the Bloomberg Philanthropies Global Forum in New York, Jane Goodall offered what became her final, unedited public reflection.

She looked back on her life from a place of deep perspective, urging people to recognize that every person has a purpose, even if it takes time to discover. Every life, she said, carries weight. Each day offers a chance to shape the world, and small choices, multiplied across millions, can create extraordinary change.

Jane reminded her audience that humans are inseparable from the natural world. Even in a time of ecological loss, she insisted that they must never abandon hope. Without hope, people stop trying; with it, they can still protect what remains beautiful and safeguard the Earth for future generations.

She spoke of her belief that consciousness continues beyond death, not as a secret she could reveal, but as a truth she now felt certain of. What we do in this life, she suggested, shapes whatever comes next.

---

## Influencing Modern Science

Jane's discoveries at Gombe did far more than redefine primatology. Her patient, immersive approach and the behaviors she revealed reshaped thinking across a wide range of scientific disciplines.

In psychology, her observations of emotion, attachment, conflict, and reconciliation in chimpanzees challenged long-held assumptions about the uniqueness of human feeling. Her work helped shift the field toward a deeper appreciation of animal cognition, empathy, and social complexity.

In anthropology, Jane's findings blurred the boundary between human and nonhuman cultures. Jane's findings blurred the boundary between human and nonhuman cultures. Anthropologists must reconsider the definition and emergence of "culture" in social groups because of toolmaking, hunting cooperation, and community-specific traditions.

In neuroscience, her documentation of learning, memory, and problem-solving in wild chimpanzees provided real-world evidence of sophisticated cognitive abilities. These insights helped inspire comparative studies of brain evolution, communication, and the neural foundations of social behavior.

In animal welfare, Jane's insistence that chimpanzees have personalities, emotions, and rich inner lives transformed global standards of care. Her work contributed to the movement away from invasive research and toward ethical treatment, sanctuary care, and legal protections for great apes.

In conservation biology, Jane's long-term study revealed how fragile chimpanzee populations are when faced with habitat loss and human pressure. Her later work through the Jane Goodall Institute helped pioneer community-centered conservation, linking wildlife protection with education, local leadership, and sustainable development.

In evolutionary biology, her discoveries illuminated the deep continuities between humans and other primates. Tool use, cooperation, warfare, maternal teaching, and cultural variation all provided powerful evidence for shared evolutionary roots and helped refine theories about how human behavior evolved.

In cognitive science, Jane's approach, observing individuals over years, not moments, revealed minds capable of planning, imitation, innovation, and social learning. Her work helped establish the study of animal cognition as a rigorous scientific field.

Across these disciplines, Jane's legacy is the same: she expanded the scientific imagination, showing that understanding other species is essential to understanding ourselves.

Jane's research helped establish long-term field observation as one of the most powerful tools in behavioral science. By following the same individuals over decades rather than weeks, she revealed patterns of development, learning, aging, and social change that short-term studies could never detect.

## Why She Matters Today

Jane's influence reaches far beyond the forests of Gombe. Her discoveries reshaped our understanding of what it means to be human, and her global advocacy continues to guide conservation, climate action, and youth empowerment.

Jane's early observations, such as chimpanzees using tools, hunting cooperatively, and expressing complex emotions, compelled science to reconsider long-held assumptions about the boundary between humans and other animals. Modern primatology, animal cognition research, and ethical standards for working with wildlife all trace their foundations to her fieldwork. The Gombe study has continued for over 65 years, making it the longest-running study of wild chimpanzees.

Through the Jane Goodall Institute, she pioneered community-centered conservation, which involves protecting ecosystems by supporting the people who live alongside them. This model is now used worldwide, influencing how governments, NGOs, and scientists approach habitat restoration, sustainable livelihoods, and wildlife protection.

Her Roots & Shoots program, active in over 60 countries, encourages young people to take practical action in their communities. It has become one of the most influential youth-led environmental movements, shaping a generation of leaders who see conservation, human rights, and peace as interconnected.

Jane is one of the world's most trusted voices on climate change and biodiversity loss. Her message, that hope is a responsibility, not a luxury, guides global conversations about how individuals, communities, and nations can protect the planet. Policymakers, scientists, and activists continue to draw on her moral clarity and scientific authority.

By sitting quietly beneath the trees of Gombe and simply watching, Jane changed forever how humanity sees its closest relatives and, perhaps, how we see ourselves. Her work continues to transform our understanding of animals and inspire global action to protect the natural world.

## Timeline

| Year | Event |
|---|---|
| 1934 | Born in London |
| 1957 | Traveled to Kenya; met Louis Leakey |
| 1960 | Arrived at Gombe; discovered termite fishing/tool use; observed hunting and meat-eating. |
| 1961 | Observed food sharing and cooperation |
| 1962 | Began PhD at Cambridge |
| 1964 | Married Hugo van Lawick |

| | |
|---|---|
| 1965 | Awarded PhD |
| 1967 | Gombe Stream Research Center established; Grub born |
| 1968 | Gombe becomes a National Park |
| 1971 | Published *In the Shadow of Man* |
| 1974 | Divorced Hugo van Lawick |
| 1975 | Married Derek Bryceson |
| 1977 | Founded the Jane Goodall Institute |
| 1980 | Derek Bryceson died |
| 1986 | Understanding Chimpanzees Conference |
| 1990 | Published *Through a Window* |
| 1991 | Launched Roots & Shoots |
| 2002 | UN Messenger of Peace |
| 2025 | Awarded the Presidential Medal of Freedom; died aged 91 |

"Every individual matters. Every individual has a role to play. Every individual makes a difference."

—*Jane Goodall*

# ADA LOVELACE

(1815-1852)

## Mathematician and Writer

Envisioned the first computer algorithm and helped shape the future of computing.

Original portrait illustration by Osmo, based on a historical reference photograph (c. 1843)

"The more I study, the more insatiable
do I feel my genius for it to be."
—*Ada Lovelace*

Ada Lovelace grew up in a world that celebrated poetry and propriety, but the logic and possibility of numbers drew her in. From an early age, she saw patterns where others saw limits. Her extraordinary ability to combine imagination with mathematics would lead her to envision the world's first computer algorithm and the future of programmable machines, more than a century before electronic computers existed.

## Childhood

Ada Lovelace was born Augusta Ada Byron on December 10, 1815, in London, England. She was the one and only daughter of the well-known romantic poet Lord George Gordon Byron and his wife, Anne Isabella Milbanke Byron, who went by Annabella.

Her parents could not have been more different. Byron was brilliant, emotional, and unpredictable, celebrated for his poetry but also notorious for his affairs and turbulent behavior. Annabella was deeply religious, disciplined, and highly educated in mathematics. Byron once nicknamed her the "Princess of Parallelograms" because of her analytical mind, though their relationship later deteriorated so badly that he mockingly called her a "mathematical Medea."

Their marriage, which began in January 1815, was unhappy from the start. Some believed Byron had married Annabella for financial reasons, while Annabella believed it was her duty to guide and reform him. Neither hope came true.

## A Marriage Falls Apart

Ada's birth did little to improve the situation. Byron had hoped for a son and reportedly reacted with disappointment. The already fragile marriage quickly unraveled under the weight of Byron's erratic behavior, infidelities, and mounting debts.

Soon after Ada's birth, Byron's half-sister, Augusta Leigh, warned Annabella that his behavior was becoming increasingly unstable and unsafe. Rumors of an inappropriate relationship between Byron and Augusta Leigh only deepened Annabella's distress.

In January 1816, when Ada was just five weeks old, Annabella made a decisive choice. She left Byron and took her infant daughter away from London, seeking refuge at her parents' country estate in Leicestershire. Shortly afterward, she pursued a legal separation.

By April 1816, the marriage was officially over. Annabella received sole custody of Ada. Byron left England forever and never saw his daughter again.

---

**HISTORICAL SPOTLIGHT**

## Lord Byron (1788–1824)

Lord Byron was one of the most influential poets of the Romantic era, known for his lyrical intensity, sharp wit, and rebellious spirit. Born George Gordon Byron in London, he rose to fame with Childe Harold's Pilgrimage and later created the archetypal "Byronic hero," a brooding, brilliant, emotionally haunted figure that shaped European literature for generations. Byron's life was as dramatic as his verse: marked by scandal, political passion, and restless travel across the Mediterranean. Known for his eccentric behavior, Byron kept a tame bear in his rooms at Cambridge, lived among poets in Italy with five peacocks wandering the grand staircase, and even wrote a grammar book of Armenian. He championed liberal causes, ultimately dying in Greece while supporting its fight for independence. However, they denied interment of his body in Westminster Abbey due to "questionable morality." Byron's legacy endures as both a poet and a cultural icon.

---

**HISTORICAL SPOTLIGHT**

## Annabella Milbanke (1792–1860)

Annabella Milbanke, later Lady Byron, was a highly educated aristocrat known for her sharp intellect, disciplined temperament, and deep commitment to social reform. Her parents raised her in a progressive household, where she excelled in mathematics. Lord Byron, her future husband, nicknamed her the "Princess of Parallelograms." Their brief and turbulent marriage ended within a year, leaving Annabella to raise their daughter, Ada Lovelace, alone. Determined to give Ada a stable, intellectually rigorous upbringing, she emphasized mathematics, logic, and moral discipline, an educational path that helped shape Ada's extraordinary analytical mind.

---

## A Father Lost

Ada grew up without ever knowing her father. Although Byron kept a picture of her and occasionally inquired about her well-being, he played no role in her life. Before his death, he was told that Ada was "not devoid of imagination, but chiefly occupied with mechanical ingenuity," a description that would prove remarkably accurate.

Byron left England under a cloud of scandal and debt, eventually traveling to Greece, where he supported the Greek struggle for independence. He died there in 1824, when Ada was just eight years old.

## A Carefully Controlled Childhood

Annabella was determined that Ada would not grow up to resemble her father. She feared that imagination and emotional intensity could lead to instability, so she designed her daughter's upbringing with strict intention. Instead of literature or poetry, Ada's education focused heavily on mathematics, logic and reasoning, as well as science and discipline.

Annabella believed that a structured and analytical education would protect Ada from what she saw as Byron's "dangerous" traits. She monitored her daughter closely, shaping nearly every aspect of her development.

Her mother even shielded Ada from her father's image. In their home, Annabella hid a portrait of Byron behind a curtain and forbade Ada from looking at it. On one occasion, when she asked to see it, she became so overwhelmed that she had to be taken from the room in tears.

Although she never knew him, Byron's legacy followed Ada everywhere. His fame, and the scandals surrounding him, made her a subject of public curiosity from birth.

Ada also had half-siblings she would never meet, including Allegra Byron, Byron's daughter with Claire Clairmont, who tragically died young. These distant connections only added to the complex and fragmented family history that surrounded her childhood.

## The Foundations of Genius

By the time she was a young child, a brilliant but absent father, a strict and controlling mother, and a carefully engineered education had already shaped Ada's life, focused on science

What Annabella intended as protection would ultimately have an unexpected effect. By encouraging logic and mathematics while suppressing imagination, she created the perfect conditions for Ada to develop something rare: a mind that could combine creativity with calculation. This unique way of thinking would later allow Ada Lovelace to imagine something no one else could see: the future of computing.

Removed from her father's influence and raised under her mother's strict guidance, Ada entered a childhood defined by structure, discipline, and education.

## Early Education

Despite the scandal surrounding her parents' separation, Anabella did all she could to shield her daughter from its effects. Well-educated herself, she strongly believed that Ada should receive a rigorous and structured education.

Ada's grandmother arranged for a personal governess, and from an early age, her governess taught Ada a wide range of subjects, including history, literature, languages, geography, music, chemistry, sewing, mathematics, and horse riding. However, Ada lived a largely isolated life at Kirkby Mallory Hall. Ada spent most of her time with governesses, tutors, and her beloved pet cat, Mrs. Puff.

Even as a child, Ada found patterns, puzzles, and machines captivating. Her education, guided by her mother and private tutors, introduced her to mathematics, music, and languages, laying the foundation for the analytical mind that would later envision the Analytical Engine.

Anabella was a forceful and highly controlling parent who implemented what she called her "system" of education. Although she was often physically absent, she remained deeply involved in overseeing Ada's studies, carefully selecting and replacing tutors as needed. Like many aristocratic children of the time, Ada received a home education and benefited from access to some of the finest tutors available.

A skilled mathematician herself, Anabella believed mathematics would ground Ada's mind. She deliberately structured Ada's education to emphasize logic and scientific thinking while avoiding poetry and imaginative literature. Annabella feared that too much imagination might lead Ada toward the same instability she believed had characterized her father.

As a result, from the age of four, Ada received an unusually strong education in mathematics and science, subjects rarely emphasized for girls at the time. Alongside this, she learned music, drawing, and languages, eventually becoming fluent in French, as well as acquiring the social accomplishments expected of a young woman of her class.

Annabella's determination to maintain this focus was unwavering. When tutors did not follow her strict educational philosophy, Annabella quickly replaced them. One example occurred when Ada, around six years old, studied under Miss Lamont. Despite her mother's emphasis on mathematics, Ada preferred geography and only studied arithmetic to please her. When

Annabella discovered this, she insisted on replacing one of Ada's geography lessons with arithmetic and shortly afterward dismissed Miss Lamont altogether.

The emphasis on mathematics continued throughout Ada's childhood, although it was not always easy for her. Between the ages of eight and twelve, she studied under William Frend, a trusted friend of Annabella who had also been her tutor. While Frend provided a firm foundation in mathematics, Ada sometimes struggled. At ten years old, she confessed to her mother that she was "a little afraid of theorems," but promised she'd "do [her] best."

Fearing that Ada might inherit her father's temperament, Annabella closely monitored her daughter's emotional state. Referring to herself as "the hen" watching over her chick, she maintained strict control over Ada's behavior and studies. Some family members worried Ada was being pushed too hard, but Annabella dismissed these concerns.

Discipline was often strict. Sometimes, Ada had to lie motionless for extended periods to develop self-control, or she was placed in solitary confinement. She also had to write apologies for perceived shortcomings, such as this one: "I, Ada, have not done the Notes very well, but I'll try to do it better tomorrow."

The relationship between Ada and her mother was not close. While living at Kirkby Mallory Hall, Ada spent much of her time with her grandmother, Lady Milbanke. Although Annabella presented herself publicly as a devoted mother, partly to secure custody, she was frequently absent, often pursuing various health treatments.

Lord Byron noted her absences even. In a letter dated March 1, 1816, he wrote: "They tell me young [Ada] is well and shows marvellous indications of acquaintance with her nurse and her grandmother, it is perhaps time she should begin to recognise another of her relations."

The remark, however, carries a sense of irony, given his own absence from Ada's life.

Annabella also relied on close acquaintances to observe Ada for any signs of undesirable behavior. Ada later referred to these individuals as the "Furies," claiming they exaggerated or fabricated reports about her. In addition, Annabella enlisted Miss Arabella Lawrence to help correct what she described as Ada's "argumentative disposition."

At eleven years old, Ada accompanied her mother on a Grand Tour of Europe. They departed for the continent in June 1826 and returned to England fifteen months later, in the fall of 1827.

## Imagining Flyology

After returning from Europe, Annabella rented a country house called Bifrons, about sixty miles from London. As was often the case, she was away for extended periods, leaving Ada free to pursue her own interests. It was during this time that Ada's fascination with flight took shape.

From an early age, Ada had shown a strong interest in machines. She spent hours studying diagrams of the new inventions of the Industrial Revolution, which filled the scientific magazines of the time. By the age of seven, she had already become captivated by the idea of designing a flying machine. She imagined a device powered by a steam engine, capable of moving large wings and lifting a person into the air.

By 1828, at just twelve years old, Ada had developed these ideas further. In a letter to her mother, she described her vision: "I have got a scheme… to make a thing in the form of a horse with a steam engine in the inside, so contrived as to move an immense pair of wings… in such a manner as to carry it up into the air while a person sits on its back."

Determined to bring her idea to life, Ada carefully studied the anatomy of birds and considered the properties of different materials. She constructed a set of wings made from wire, covered with paper and feathers, and worked through calculations related to lift and propulsion.

Her enthusiasm for the project was intense. She even began writing a book, which she titled *Flyology*, where she documented and illustrated her designs, including plans for a steam-powered mechanical flying horse.

However, her growing focus on this imaginative project concerned her mother. Believing that Ada was drifting too far into creativity, traits she associated with Lord Byron, Annabella intervened. She reprimanded Ada for neglecting her formal studies and responded by further intensifying her mathematical education, even hiring an expensive tutor to ensure Ada remained focused on logic and discipline.

Despite this, Ada's work on *Flyology* revealed an important aspect of her character: a rare ability to combine imagination with scientific reasoning. This blend of creativity and analytical thinking would later define her approach to mathematics and invention.

Years later, her friend Charles Babbage would affectionately refer to her as "Lady Fairy," a nod to the imaginative spirit that remained a part of her throughout her life.

## The Teenage Years

Ada's early years were tough, marked by hard studies and frequent bouts of illness. From a young age, she suffered from frequent health problems, including severe migraines that affected her vision by the age of eight.

In 1829, at thirteen years old, Ada contracted a serious case of measles. The illness left her partially paralyzed and in fragile health. She spent over a year confined to bed, a period of prolonged recovery that likely delayed her physical development.

Even after regaining some strength, Ada's mobility remained limited. By her early teens, people often saw her walking with the aid of crutches. These ongoing health challenges would continue to affect her throughout her life.

Despite her health struggles, Ada's education continued with intensity. Her studies extended far beyond basic reading and writing, encompassing advanced arithmetic, geometry, algebra, astronomy, and music. These subjects would later form the foundation for her insights into mechanical calculation.

Although Annabella had carefully structured her education to suppress imaginative tendencies, Ada did not abandon her creative instincts. In her own words, she aspired to be "an analyst and a metaphysician." She came to view mathematics not as rigid or limiting, but as a field rich in imagination and possibility. Reflecting this perspective, she has said, "If you can't give me poetry, at least give me poetical science."

Around the age of thirteen, Ada and her mother moved to Branch Hill House in Hampstead, England. During this time, she studied under Dr. William King. However, he had limited mathematical interest and stated that he learned the subject more by reading than by practicing.

At sixteen, Ada moved again with her mother, this time to Fordhook in Ealing. During the following two years, her education intensified further, continuing to build the analytical foundation that would later define her work.

## A Forbidden Romance

In 1832, at seventeen years old, Ada became involved in a relationship with one of her tutors, William Turner, who had been teaching her shorthand. Her mother strongly disapproved of the relationship.

In response, Ada ran away to the home of Turner and his family, intending to elope. However, the situation quickly came to light. Recognizing Ada, Turner's relatives contacted her family, who soon returned to her home.

Annabella acted swiftly to suppress the incident, ensuring that it did not become a public scandal. She kept the matter private, which preserved Ada's reputation within society.

## London Scientific Society

The following year marked a turning point in Ada's life. In 1833, William Frend, who had previously taught both Ada and her mother, introduced Ada to Mary Somerville.

Annabella admired Mary and encouraged the connection. Mary, often referred to as the "Queen of Nineteenth-Century Science," was an accomplished mathematician and astronomer. The Royal Astronomical Society, despite the limited participation of women in scientific circles at the time, included her as one of its first female members. Widely respected, she served as a model of what a serious female intellectual could achieve. Ada would develop a close and lasting relationship with her.

By this time, Ada had grown into a complex and strong-willed young woman. In contrast to the expectations of Victorian society, which favored modesty, obedience, and restraint in young ladies, Ada was independent, intellectually curious, and sometimes contradictory. She could be both charismatic and forceful, confident in her ideas yet occasionally uncertain of herself.

Annabella, despite often missing Ada's childhood, formally introduced her daughter into London society. This "coming out" allowed Ada to enter influential social and intellectual circles, where she encountered many of the leading thinkers of the day.

Through her mother's connections and her growing friendship with Mary, Ada began attending gatherings within the London scientific society, a network of salons, soirees, and intellectual meetings. There, she met prominent figures across disciplines, including Augustus De Morgan, Andrew Crosse, David Brewster, Charles Wheatstone, Michael Faraday, Charles Darwin, and Charles Dickens.

During this period, Ada began corresponding with leading scientific minds, not merely as an observer, but as someone engaged in an intellectual exchange with ideas of her own. The Royal Society rarely admitted women, and universities effectively closed their doors to them. Yet Ada, an earl's daughter and later Countess, could enter spaces most women could not, attending salons, corresponding with leading scientists, and observing machines in action, a privilege that shaped her extraordinary trajectory.

Despite her mother's efforts to distance her from her father's influence, Ada maintained a deep reverence for Lord Byron throughout her life. She shared aspects of his imaginative nature and often described her own work as "poetical science," believing that imagination and analytical thought were not opposites, but complementary forces. She also regarded metaphysics as essential to exploring what she described as "the unseen worlds around us."

During this time in 1833, someone introduced Ada to Charles Babbage at a dinner party, a meeting that would prove one of the most significant of her life.

---

**HISTORICAL SPOTLIGHT**

## Mary Somerville (1780–1872)

Mary Somerville was one of the most respected scientific thinkers of the 19th century, a mathematician whose clarity and discipline shaped Ada Lovelace's intellectual world. Born in Scotland and largely self-taught, she became internationally known for her translation and explanation of Laplace's Mécanique Céleste, a work so lucid that it opened advanced celestial mechanics to English readers for the first time. By the early 1830s, her London salon had become a gathering place for astronomers, mathematicians, and instrument makers, offering Ada a rare doorway into scientific society. Somerville's mentorship was gentle but rigorous; she encouraged Ada's curiosity, guided her studies, and ultimately introduced her to Charles Babbage, an encounter that changed the course of Ada's life and the history of computing.

---

# Charles Babbage

On May 10, 1833, at 17 years old, Ada was formally presented at court to King William IV and Queen Adelaide. Her intelligence and "brilliant mind" impressed everyone, not just her social presence, and the court quickly labeled her the "popular belle of the season."

Yet it was an event less than a month later that would prove far more significant. On June 5, 1833, Ada attended a gathering with her mother and Mary at the home of Charles, at 1 Dorset Street in Marylebone, London.

At forty-one years old, twenty-four years Ada's senior, Charles was already an established mathematician, philosopher, and inventor. The son of a wealthy banker, he received his education at Cambridge University and later assumed the position of Lucasian Professor of Mathematics, a role also held by Isaac Newton and Stephen Hawking. He was also a founding member of both the Astronomical Society (1820) and the Statistical Society (1834).

However, tragedy marked Charles's personal life. In 1827, his father, wife, and two of his children died within a short period. Following these losses, he devoted himself even more fully to his work. His father's inheritance, valued at approximately £100,000, provided him with financial independence, allowing him to pursue his scientific interests without constraint.

When Ada met Charles, she was already something of a society figure, known in part because of her father's notoriety. Charles may have found her presence especially striking, as his own daughter, had she lived, would have been of a similar age.

At first, Charles underestimated Ada, attempting to engage her with one of his usual party attractions, a mechanical automaton known as the "Silver Lady." However, Ada's interest quickly moved beyond novelty. She engaged deeply in conversation with both Charles and Mary, demonstrating a genuine curiosity about scientific and mathematical ideas.

During the evening, Charles described his work on an automated calculating machine known as the Difference Engine, as well as the difficulties he faced securing government funding for the project. Impressed by Ada's understanding and insight, he invited her and her mother to return for a private demonstration of the machine on June 17, 1833. The demonstration enthralled Ada. As later recalled by Sophia Frend, daughter of William Frend: "...young as she was, she understood its working and saw the great beauty of the invention."

By the spring of 1834, Ada and her mother had formed a growing friendship with Charles. Together with Mary, they began attending his well-known Saturday evening gatherings, which brought together many of the leading intellectual figures of the time.

Over the years that followed, Ada and Charles developed a close intellectual partnership. They corresponded regularly with mathematical ideas, forming a friendship that would endure throughout their lives. In the early years, however, their interactions remained largely within the boundaries set by society, often taking place in the presence of Annabella.

---

**HISTORICAL SPOTLIGHT**

## Charles Babbage (1791–1871)

Charles Babbage grew up in a prosperous banking family and showed early mathematical talent, eventually studying at Cambridge. Frustrated by the university's outdated methods, he joined John Herschel (son of the discoverer of Uranus) and George Peacock (later a pioneer in abstract algebra) in founding the Analytical Society (which later became the Cambridge Philosophical Society), which pushed to modernize British mathematics like replacing Newton's ("British") dot-based notation for calculus with Leibniz's ("Continental") function-based one. After graduating in 1814, Charles began building a reputation through lectures and modest mathematical papers. In 1827, Charles's father died, leaving him about £100K, or perhaps $14 million today, setting Charles up financially for the rest of his life. The same year, though, his wife died. She had eight children with him, but only three survived adulthood. Feeling low after his wife died,

Charles visited Europe. He was so taken with the science there that he penned a book, Reflections on the Decline of Science in England.

## The Difference Engine

Charles had long written competent, though not revolutionary, papers on mathematics and economics. But a trip to France in 1819, where he observed large-scale table-making projects, inspired a remarkable idea that the tables "had been made by steam." Charles later described the moment that sparked his vision of a mechanical calculator: "I was sitting in the rooms of the Analytical Society in Cambridge with a book of logarithms open before me. Another member, seeing me half asleep, called out, 'Well, Babbage, what are you dreaming about?' To which I replied, 'I am thinking that these tables might be calculated by machinery.'"

By the summer of 1821, he had a clear vision for a cogwheel-based calculating machine capable of producing mathematical tables reliably and automatically. Such tables were of critical military and commercial importance, used for navigation, construction, and a wide range of practical calculations.

Before mechanization, human clerks calculated tables, themselves called "computers," who performed arithmetic mentally and recorded their work. Although skilled, they often made errors, and mistakes frequently spread through the calculations that depended on their work.

In 1827, Charles published a corrected table of logarithms, calculated entirely by hand, which became a standard reference for nearly a century. Printed on yellow paper to reduce user error, the table appeared in *A Table of the Logarithms of the Natural Numbers from 1 to 108,000* and highlighted the frequency of mistakes in existing tables. It also served as evidence of the urgent need for a mechanical calculator: the Difference Engine.

The Difference Engine was designed to compute any polynomial up to a certain degree using the method of differences. It could automatically step through values and print results, eliminating human error. Recognized for its innovation, Charles received a medal from the Astronomical Society he had co-founded, and in 1823, the British government agreed to fund the project with £17,000 (roughly $2 million today).

Charles was meticulous in drafting plans and conducting experiments in his machine shop, yet he struggled with managing the engineers he hired. Despite advances in precision tooling, he could not bring the full machine to completion. By 1832, however, he had built a partial prototype: a working model one-seventh the size of the planned engine, the very device Ada would see in 1833.

This prototype stood about two feet high, hand-cranked, and composed of roughly 2,000 brass parts. Vertical columns of cogwheels represented numbers, and turning a handle allowed the gears to perform calculations. Though limited to addition and subtraction, the machine could generate complex tables automatically.

Charles never completed a fully operational Difference Engine. Yet the design revealed a groundbreaking concept: controlling computational steps mechanically, with the potential for data storage. He even incorporated the idea of punched cards, inspired by Jacquard looms (1801), to encode sequences of operations.

Annabella referred to the machine as a "thinking machine," noting that it could "raise several Nos. to the 2nd & 3rd powers, and extract the root of a Quadratic Equation." Ada immediately grasped the underlying mathematics and began conceptualizing how the machine could be "programmed" to perform tasks systematically, a precursor to modern computing.

Intrigued by the prototype, Ada corresponded with Charles about its potential and shared her own mathematical ideas. Their letters, spanning from June 10, 1835, to August 12, 1852, reveal a dynamic intellectual partnership. In 1839, Charles encouraged her, writing: "I think your taste for mathematics is so decided that it ought not to be checked."

Charles would later call Ada the "Enchantress of Numbers," praising her ability to grasp abstract concepts and stating: "...she has thrown her magical spell around the most abstract of sciences and has grasped it with a force which few masculine intellects… could have exerted over it."

Despite Ada's insight, the Difference Engine never advanced beyond preliminary prototypes. High costs, complex components, and managerial challenges hindered completion. Efforts to secure further government funding failed, even after Charles proposed a simplified Difference Engine No. 2, which required 8,000 parts instead of 25,000.

Ultimately, Charles shifted his focus to an even more ambitious design: the Analytical Engine, a general-purpose calculating machine that would fully realize the potential of his earlier innovations.

---

**SCIENCE SPOTLIGHT**

### Jacquard's Loom and Punch Cards (1804)

In 1804, the French weaver Joseph Marie Jacquard introduced a loom that could weave complex patterns automatically using a sequence of punched cards. Each card encoded a row of the design. A hole signaled the lifting of a thread. By chaining cards together, the loom could "remember" and repeat intricate motifs with remarkable precision. This breakthrough transformed the textile industry, reducing errors and allowing designs that once required master weavers. More importantly for computing history, Jacquard's punched card system became a conceptual ancestor to programmable machines. Jacquard's insight that information could be stored, ordered, and executed mechanically directly influenced Charles Babbage's Analytical Engine and, later, Hollerith's tabulating machines.

---

## Marriage to William King

During a philanthropic tour of northern England's mills in 1834, Ada accompanied her mother and became fascinated by the advanced machinery she found. On the return journey, she taught some mathematics to the daughters of one of her mother's friends, continuing the lessons by mail. Ada joked that this could be 'the commencement of "A Sentimental Mathematical Correspondence carried on for years between two ladies of rank," to be hereafter published, no doubt, for the edification of mankind, or womankind.'" Though not highly sophisticated, her instructions were clear and precise, complete with admonitions such as, "You should never select an indirect proof when a direct one can be given."

Later that year, Annabella began seeking a suitable suitor for Ada, insisting she marry an aristocrat. In the spring of 1835, Lady Byron introduced nineteen-year-old Ada to 30-year-old Lord William King-Noel, an amiable but not especially brilliant aristocrat. Ada quickly took the lead in the relationship.

William, Mary's son's friend, attended Eton and Cambridge and worked as a civil servant, spending time on the Greek islands. In June 1835, Ada wrote to him: "I am going this evening to my friend Mrs. Somerville's to stay the night. She has kindly offered to take me to a concert, which my love of music could not resist."

After a brief courtship, they married on July 8, 1835. William proved devoted, sharing Ada's love of horses and admiring her intellect. He reportedly remarked, "What a general you would make." Ada, however, was determined not to let marriage slow her work. Shortly after her wedding, she wrote to Mary: "I now read Mathematics every day and am occupied in Trigonometry and in preliminaries to Cubic and Biquadratic Equations. So, you see that matrimony has by no means lessened my taste for these pursuits, nor my determination to carry them on."

Through marriage, Ada gained considerable wealth: estates at Ockham Park in Surrey, a summer retreat in Somerset, a Scottish estate beside Loch Torridon, and a large London residence. This financial security allowed her to pursue her intellectual interests freely, unconstrained by conventional obligations.

Although Ada engaged in an extramarital affair, William forgave her. On May 12, 1836, the couple welcomed their first child, Byron King-Noel. A daughter, Anne Isabella, was born on September 22, 1837. During this period, Ada continued to struggle with health issues, including breathing and digestive problems, and experienced rheumatic attacks following childbirth.

In 1837, Queen Victoria ascended the throne, and as a prominent member of high society, Ada met her. In 1838, the family revived the extinct line of Baron Lovelace, from which Ada descended. The British government recognized William's service by creating him the first Earl of Lovelace and Viscount of Ockham, and Ada then became Lady Ada King, Countess of Lovelace, leading to her lasting title, Ada Lovelace.

A second son, Ralph Gordon King-Milbanke, was born on July 2, 1839. Shortly afterward, Ada recommitted to mathematics. In London, she sought a "mathematical instructor" and requested anonymity to avoid society gossip.

Over the following years, Ada balanced her roles as mother and household manager with pursuits in horse riding, music, and mathematics, including studies in spherical trigonometry. She and her husband socialized with prominent figures such as Michael Faraday and Charles Dickens.

Ada's relationship with her mother remained complex. Outwardly respectful, she often found Annabella controlling and manipulative, with family fortunes largely directed by her mother, hardly opposed by William. Annabella's frequent claims of failing health and constant critiques of Ada's child-rearing, household management, and social conduct were a continual source of tension. Yet Ada's ambition remained undiminished. On February 6, 1841, she wrote to her mother: "I believe myself to possess a most singular combination of qualities exactly fitted to make me preeminently a discoverer of the hidden realities of nature."

She described an "insatiable and restless energy" and a newfound sense of purpose, feeling less "secretive and suspicious" of her mother.

Shortly after, Annabella revealed a family secret: before Ada's birth, Lord Byron had fathered a child with his half-sister. Though not illegal in England, it was scandalous. The revelation deeply affected Ada, temporarily derailing her mathematical work.

Encouraged by her husband, Ada returned to mathematics by late 1842. She contributed to William's writings on agriculture, explored mesmerism, and contemplated whether the operations of the brain might follow mathematical laws, a "calculus of the nervous system." William also supported her intellectual pursuits by copying articles for her in libraries, as libraries often denied women direct access.

Through wealth, family support, and determination, Ada navigated the complexities of high society, motherhood, and health challenges, all while continuing to cultivate her mathematical genius.

## Augustus De Morgan

Mathematics was central to Ada's life; she once wrote that it "constitutes the language through which alone we can adequately express the great facts of the natural world."

Ada first met one of the leading intellectuals of her time, the mathematician and logician Augustus De Morgan, through Mary as part of the London Scientific Society in 1833. As William Frend, Ada's longtime tutor, grew older and fell behind in contemporary mathematical developments, Ada outgrew his instruction. Therefore, Frend recommended De Morgan to her for more advanced studies.

Initially, Ada continued to work with Mary. But after the birth of her third child and with encouragement from Charles, she renewed her serious focus on mathematics and asked De Morgan to tutor her in advanced, university-level mathematics.

De Morgan, the first professor of mathematics at London University (now University College London) and a prominent mathematical educator, was ideally suited as a guide for her. Since universities barred women from attending, Ada began a correspondence course with him in 1840, based on the material he taught his advanced UCL mathematics students. De Morgan organized his instruction around fundamental principles and praised Ada as a thinker whose ability was "utterly out of the common way," capable of grasping the "real difficulties of first principles."

Their letters covered topics such as algebra and calculus, and Ada speculated in a forward-looking way about the potential of mathematics and computing, for example, a "calculus of the nervous system" or using mathematics to investigate music. During the correspondence, De Morgan gently challenged Ada to address any gaps in her knowledge, ultimately strengthening the foundation that would support her later work with Charles and her seminal *Notes on the Analytical Engine*. Ada, however, was not shy about questioning or debating controversial ideas with her teacher.

De Morgan was so impressed by Ada's abilities that he wrote to her mother and declared that her mastery of mathematics was such that "if she were a man, they would certainly have made him an original mathematical investigator, perhaps of first-rate eminence." He also shared his admiration with colleagues, including Maria Gaetana Agnesi, the Italian author of a calculus textbook, rating Ada's talent even above that of Mary.

## The Analytical Engine

When Charles failed to secure government support for the second version of the Difference Engine, he was undeterred. By 1834, he had already begun envisioning a larger, more powerful machine: the Analytical Engine. Babbage conceived the Analytical Engine as a general-purpose calculating machine, effectively the world's first digital computer, unlike the Difference Engine, which was designed solely to calculate mathematical tables.

Charles imagined that the Analytical Engine could automatically produce mathematical tables, either printing them or plotting them on a plate. While humans would remain primary users, he envisioned libraries of pre-computed cards that could provide machine-readable data.

Like the Difference Engine, the Analytical Engine relied on cogwheels, gears, and levers for its calculations. Charles's design plans grew increasingly detailed, eventually comprising over 300 drawings and 2,200 notations. Included in the machine was a central processing unit, which Charles called the Mill, expected to rise fifteen feet (4.5 m) high, as well as a memory unit or Store, along with a printer, card punch, and graph plotter. The engine also incorporated safety mechanisms to warn operators of errors.

The Analytical Engine also employed the punched-card technology that Joseph-Marie Jacquard had developed for his pattern-weaving looms. These cards provided both data and instructions for the machine, and anyone could use them in any order rather than sequentially. The machine could also repeat a deck of cards, effectively executing loops, which made it far more flexible than the Difference Engine, able to follow instructions and execute sequences rather than performing a single type of calculation.

Charles further introduced the concept of conditional control transfer, allowing the engine to pause mid-calculation and select between alternative steps based on values it had already determined. This early form of decision-making, or branching, gave the machine the ability to execute conditional instructions.

The design of the Store envisioned holding one hundred 50-digit numbers, requiring a space of about twenty feet (6 m) in length. Charles estimated it would take approximately three minutes to multiply two 20-digit numbers and that the machine would need to be powered by steam because of its size.

Ada was even more captivated by Charles's vision of the Analytical Engine than she had been by the Difference Engine. Charles later noted that he had met no one who grasped the machine's potential as quickly or as completely as Ada did. She would act as its key interpreter, translating its complex design into concepts accessible to others.

Like the Difference Engine, Charles only built partial components of the Analytical Engine to test feasibility, including:

- The "Addition and Carry" mechanism
- Sections of the Mill's gear trains
- Trial assemblies for the Store's digit wheels
- Card-reading experiments using adapted Jacquard mechanisms

These prototypes showed that the Engine's logic was mechanically sound. However, Charles never completed the Analytical Engine. It was vastly more complex than the Difference Engine, requiring thousands of precision-machined parts and funds far beyond Charles's means. Parliamentary sponsors, frustrated by the incomplete Difference Engine, refused to fund the new project, prompting Charles to seek support abroad.

**SCIENCE SPOTLIGHT**

## How the Analytical Engine Worked

Charles Babbage's Analytical Engine was the first design to contain all the essential elements of a modern computer. At its core were two major components: the Store, which held numbers much like memory, and the Mill, which performed operations like a Central Processing Unit (CPU). Operators fed instructions and data into the machine using punched cards, adapted from Jacquard looms, which enabled the Engine to follow sequences, repeat steps, and decide based on intermediate results. A system of gears, levers, and rotating drums carried out addition, subtraction, multiplication, and division automatically. Crucially, the Engine could execute conditional branching, meaning it could change its behavior depending on the outcome of a previous calculation. This ability to combine memory, operations, and programmable control is why historians consider the Analytical Engine the first true computer.

| Analytical Engine | Modern Computer |
| --- | --- |
| Store | Memory (RAM) |
| Mill | CPU |
| Punched cards | Program |
| Printer | Output device |
| Conditional branching | If statements |
| Loops | Repeated instructions |

## The Translation Notes

In 1840, the Italian government invited Charles to give lectures on the Analytical Engine in Turin to leading mathematicians and engineers. The Italian mathematicians and engineers met his lectures, demonstrations, and mechanical drawings with exceptional enthusiasm. They produced detailed Italian analyses, which were the first substantial publications on the Analytical Engine outside Britain.

Following this visit, Luigi Menabrea, an Italian military engineer and future Prime Minister, published an article in a Swiss academic journal in October 1842 entitled *Notions sur la machine anilities de Charles Babbage* ("Elements of Charles Babbage's Analytical Machine"). Charles initially asked Mary to translate the article, which was written in French, into English. Too busy, Mary recommended Ada, who was fluent in French. Charles Wheatstone, a mutual friend, also encouraged Ada to take on the task.

Charles, recognizing how well Ada understood the machine, asked why she had not written her own paper. Ada replied that the thought had never occurred to her, and Charles suggested she add her own notes to Menabrea's article. Ada, calling herself "an Analyst (& Metaphysician)," agreed. Few women wrote for scientific publications; most who dabbled in science wrote for a nonprofessional female audience and explained concepts discovered by men. Charles's suggestion placed Ada in the unprecedented position of writing for a male scientific audience.

Ada and Charles corresponded during this period, exchanging ideas and clarifications, but Ada ultimately decided what to include in her paper. In one letter, she described the conceptual power of the Analytical Engine, essentially the idea of a general-purpose computer, writing: "It was suited for *developping [sic] and tabulating any function whatever… the engine [is] the material expression of any indefinite function of any degree of generality and complexity."

For nine months in 1842–43, Ada worked on the translation and, more significantly, expanded the article with her own set of clarifying notes, as she modestly called them. Ada's mind leaped beyond mere calculation. She thought of the Analytical Engine not just as a machine for numbers, but as one that could work with symbol sequences, foreshadowing how today's computers handle logic, language, and even music.

In 1843, when Ada was 27 years old, *Scientific Memoirs Selected from the Transactions of Foreign Academies of Science and Learned Societies* published the translation and Ada's Notes. The article was titled *Sketch of the Analytical Engine invented by Charles Babbage, with Notes by the Translator*. Ada signed only with her initials, "A.A.L.," concealing her identity, as many women did before society widely accepted female intellectuals.

Ada's Notes were extensive, three times as long as Menabrea's original eight-thousand-word article, totaling around twenty thousand words. Charles later wrote, "Their author has entered fully into almost all the very difficult and abstract questions connected with the subject." The Notes included the first documented computer program, divided into seven sections, Notes A–G.

In Note A, Ada made the first distinction between numbers and symbolic operations, realizing that a machine could manipulate symbols, not just numbers, according to rules. She explained: "The distinctive characteristic of the Analytical Engine… is the introduction into it of the principle which Jacquard devised for regulating, by means of punched cards, the most complicated patterns in the fabrication of brocaded stuffs. It is in this that the distinction between the two engines lies… The Analytical Engine weaves algebraical patterns just as the Jacquard loom weaves flowers and leaves."

She contrasted it with the Difference Engine: "The Analytical Engine is an embodying of the science of operations, constructed with peculiar reference to abstract number as the subject of those operations. The Difference Engine is the embodying of one particular and very limited set of operations…"

Ada explained the use of Operation Cards (defining the operations) and Variable Cards (defining the location of values), and introduced the concept of cycles and nested cycles, now known as loops and nested loops. In Note G, she illustrated a sequence of computations on the Analytical Engine, including a table for calculating the seventh Bernoulli number, a sequence of rational numbers named after Jacob Bernoulli. The Notes explained how the

Mill drew quantities from the Store, processed them, and returned the results according to the instructions encoded on the punched cards.

Ada's letters reveal her process and personality. In July 1843, she wrote to Charles: "I want to put in something about Bernoulli's Numbers, in one of my Notes, as an example of how an implicit function may be worked out by the engine, without having been worked out by human head & hands first... Give me the necessary data & formulae."

She expressed frustration and delight in equal measure while debugging the calculation: "My Dear Babbage. I am in much dismay at having got into so amazing a quagmire & botheration with these Numbers... I am now going out on horseback. Tant mieux."

Later, she reported success: "I have worked incessantly, & most successfully, all day. You will admire the Table & Diagram extremely... William ('Lord L.') is at this moment kindly inking it all over for me. I had to do it in pencil..."

Charles had offered to assist with the Bernoulli number algorithm, but Ada detected a major mistake in his work and returned his notes for amendment. In effect, Ada was debugging a computer program more than a century before electronic computers existed. Toward the end of the project, tensions arose; Lovelace refused to let Charles add a strong criticism of the British government, and Charles declined further involvement in the paper's organization.

Ada intentionally chose a complex example, not for simplicity, but to illustrate the power of the Analytical Engine. The method she described in Note G is now considered the first published computer program, providing a complete view of all successive changes in the machine's components during a calculation.

While Charles saw the Analytical Engine primarily as a tool for producing mathematical tables, Ada imagined broader applications. She recognized that the Engine "might act upon other things besides number," including music, if someone could express the relationships mathematically. "...Supposing, for instance, that the fundamental relations of pitched sounds in the science of harmony and of musical composition were susceptible of such expression and adaptations, the engine might compose elaborate and scientific pieces of music of any degree of complexity or extent."

She also clearly delineated the machine's limitations, emphasizing that it could not originate ideas: "The Analytical Engine has no pretensions to originate anything. It can do whatever we know how to order it to perform."

Beyond technical insight, Ada articulated the poetic and philosophical significance of Charles's invention: "This science constitutes the language through which alone we can adequately express the great facts of the natural world... A new, a vast, and a powerful language is developed for the future use of analysis... bringing the mental and the material, the theoretical and the practical, into more intimate and effective connection."

Through her Notes, Ada became the first to envision the concept of modern computing, laying the foundation for the idea that machines could go beyond calculation to perform complex symbolic operations.

NOTES BY THE TRANSLATOR.

NOTE A.—Page 671.

The particular function whose integral the Difference Engine was constructed to tabulate, is

$$\Delta^7 u_z = 0.$$

The purpose which that engine has been specially intended and adapted to fulfil, is the computation of nautical and astronomical tables. The integral of

$$\Delta^7 u_z = 0$$

being $$u_z = a + bx + cx^2 + dx^3 + ex^4 + fx^5 + gx^6,$$

the constants $a$, $b$, $c$, &c. are represented on the seven columns of discs, of which the engine consists. It can therefore tabulate *accurately* and to an *unlimited extent*, all series whose general term is comprised in the above formula; and it can also tabulate *approximatively* between *intervals of greater or less extent*, all other series which are capable of tabulation by the Method of Differences.

The Analytical Engine, on the contrary, is not merely adapted for *tabulating* the results of one particular function and of no other, but for *developing and tabulating* any function whatever. In fact the engine may be described as being the material expression of any indefinite function of any degree of generality and complexity, such as for instance,

$$F(x, y, z, \log x, \sin y, x^p, \&c.),$$

which is, it will be observed, a function of all other possible functions of any number of quantities.

In this, which we may call the *neutral* or *zero* state of the engine, it is ready to receive at any moment, by means of cards constituting a portion of its mechanism (and applied on the principle of those used in the Jacquard-loom), the impress of whatever *special* function we may desire to develope or to tabulate. These cards contain within themselves (in a manner explained in the Memoir itself, pages 677 and 678) the law of development of the particular function that may be under consideration, and they compel the mechanism to act accordingly in a certain corresponding order. One of the simplest cases would be, for example, to suppose that

$$F(x, y, z, \&c. \&c.)$$

is the particular function

$$\Delta^n u_z = 0$$

which the Difference Engine tabulates for values of $n$ only up to 7. In this case the cards would order the mechanism to go through that succession of operations which would tabulate

$$u_z = a + bx + cx^2 + \ldots . m x^{n-1},$$

where $n$ might be any number whatever.

Figure 3. A page from Ada Lovelace's 1843 *Notes on the Analytical Engine*, in which she explained how Charles Babbage's proposed machine could perform calculations beyond simple arithmetic by following programmed instructions. Many historians regard this work as the first published computer algorithm.

## SCIENCE SPOTLIGHT

# Bernoulli Numbers

Bernoulli numbers are a sequence of rational values that arise throughout mathematics, especially in the study of series, calculus, and number theory. First introduced by the Swiss mathematician Jakob Bernoulli in the late seventeenth century, they appear in formulas for sums of powers, the expansion of trigonometric functions, and the evaluation of the Riemann zeta function at even integers. Their alternating pattern and rapid growth make them both elegant and computationally demanding, one reason Ada chose them for her famous Note G example. In her table-driven program for the Analytical Engine, the machine would generate successive Bernoulli numbers step by step, demonstrating not only the Engine's capacity for complex symbolic operations but also the possibility of automated mathematical discovery.

---

**SCIENCE SPOTLIGHT**

### What Is an Algorithm?

An algorithm is a step-by-step set of instructions used to solve a problem or perform a task. Modern computer programs use algorithms to tell a computer exactly what actions to perform and in what order. People widely consider Ada Lovelace's Bernoulli number program the first published computer algorithm because it described how a machine could automatically execute a sequence of operations.

---

## The Publication Crisis and Ada's Role in the Analytical Engine Project

After Ada and Charles completed the papers, they clashed over how the work should appear in print. Charles wanted to attach a long, combative preface attacking the British government for abandoning the Difference Engine and planned to leave it unsigned so it would appear jointly authored. Ada insisted on publishing the paper without his polemic. When Charles refused to sign the required paperwork, the project nearly collapsed.

This conflict, later known as the Publication Crisis of 1843, unfolded just as Ada finished translating Menabrea's paper and expanding it with her original Notes A–G. The plan was to be published in *Taylor's Scientific Memoirs*, but Charles's refusal delayed the process. Ada held firm, and the paper ultimately appeared without his preface, leaving her Notes as the centerpiece and the most important text ever written on the Analytical Engine.

Although extraordinary, she published the Notes under her initials, "A.A.L.," and Ada did not receive official credit until 1848. Once her identity became known, some contemporaries dismissed the work more readily than if a man had written it. Still, it marked the high point of her intellectual life. Friends who knew the author behind the initials admired her achievement, even as her health declined and she struggled without a new scientific project or peers with whom she could discuss mathematics. She hoped the Notes would be the first of many publications. She considered reviewing Ohm's work on the Galvanic Series, but Charles did not encourage her. His own difficulties in securing funding for his engines increasingly frustrated him.

In August 1843, as she completed her Notes, Ada wrote a long, carefully argued letter proposing that she take on a more active role in managing the Analytical Engine project, particularly the political and interpersonal work required to secure support. She believed she could help guide the effort toward completion and prevent the setbacks caused by Charles's impulsive habits. She described this candidly to her mother, outlining the "conditions" she had set and her willingness to act as his "Whipper-in" if he agreed.

Charles rejected her proposal outright, scribbling on his copy of her letter: "Saw A.A.L. this morning and refused all the conditions." His reasons remain unclear. He valued her intellect and her ability to explain the Engine but may have been uncomfortable allowing her a formal leadership role. The disagreement did not end their friendship.

Their correspondence in the weeks that followed showed warmth, humor, and renewed camaraderie. Charles wrote about bringing drawings and papers on his next visit; Ada responded with enthusiasm. She told her mother that he was in unusually good spirits, and Charles soon wrote again, calling her the "Enchantress of Number" and "my dear and much-admired Interpreter." Ada replied playfully, teasing him for surrendering to "Fairy Guidance," and he signed his next letter "Your faithful Slave." She later described herself as the "High Priestess of Babbage's Engine."

These exchanges capture the essence of their partnership: intellectually rich, emotionally complex, and marked by mutual admiration. Even when Charles resisted her attempts to shape the project's direction, Ada's influence on the understanding and legacy of the Analytical Engine remained profound.

## Scientific Thinking

After her studies with De Morgan and the publication of her famous Notes with Charles, Ada's energies shifted to other pursuits, and she did not continue to develop a formal scientific career, a choice shaped both by circumstance and the social limitations of her era. During the 1800s, society rarely viewed women as serious scientific thinkers, and people often described Ada's talents using masculine terms. "Gentleman amateurs," lawyers, clergymen, or actuaries dominated most professional mathematics. They met at societies like the Royal Society to share ideas, but the Royal Society did not admit women.

University mathematics training focused not on producing professional mathematicians but on developing clear, rational thinking. Unlike continental Europe, Britain did not have a well-developed culture of mathematical research; the emphasis was utilitarian. Papers typically presented calculations and examples rather than formal proofs. To be a mathematician in this context meant thinking systematically about the world, grasping the power of general principles, and engaging deeply with numerical data.

While Ada maintained her interest in mathematics, she also explored philosophy and contemporary scientific discoveries, including mesmerism, hypnosis promoted by the German physician Franz Mesmer, who believed it had curative powers. Ada even considered conducting practical experiments herself and contacted Michael Faraday (1791–1867) about replicating some of his work.

Despite her mother's efforts to suppress her literary side, Ada never fully relinquished the poetic skill she inherited from her father, Lord Byron. She merged imagination and science in what she called "poetical science," emphasizing creativity as an essential tool for understanding the world: "Imagination is the Discovering Faculty, pre-eminently. It is that which penetrates into the unseen worlds around us, the worlds of Science."

Through this approach, Ada showed a rare ability to combine rigorous logic with visionary thinking, a hallmark of her enduring influence in mathematics, computing, and scientific thought.

---

**SCIENCE SPOTLIGHT**

### What is Poetical Science?

Ada Lovelace described her approach to mathematics as "poetical science." For her, imagination was not a distraction from rigorous thinking but a vital part of it. She believed that creativity allowed mathematicians to see patterns, connections, and possibilities that pure calculation alone could never reveal.

This fusion of logic and imagination shaped her work on Charles Babbage's Analytical Engine. Ada understood that the machine could manipulate both symbols and numbers, and she imagined how it might one day compose music, create art, or model complex systems. These insights came not from mathematics alone, but from her ability to think metaphorically, to imagine what a machine could become rather than what it currently was.

Ada's idea of poetical science challenged the belief that creativity and analysis belong to separate worlds. She saw discovery as a deeply imaginative act, one that requires intuition, curiosity, and the ability to think beyond the boundaries of existing knowledge. Her legacy reminds us that innovation often emerges where disciplined reasoning meets bold, imaginative vision.

---

## Turning Towards the Mind

Ada remained passionately committed to science. She called her first publication her "first-born" and saw Note A, the "science of operations," as the beginning of a larger intellectual project. Anticipating ideas that now underpin abstract computation theory, she hoped to develop a symbolic calculus describing how operations combine. She imagined a sequence of papers building on one another, much as Charles and Mary had begun their careers through translation. Yet barriers remained: as a woman, she could not access the Royal Society library, even though she had helped secure her husband's membership.

In early 1844, Ada became absorbed in Baron Karl von Reichenbach's theories of nervous forces, especially the so-called "Odic force." She drafted a critique that challenged his experimental controls, questioned the reliability of his "*sensitives*," and attempted to reinterpret his claims mathematically. Most importantly, she used his ideas to develop her own concept of a "calculus of the nervous system," an early attempt to model thought in symbolic terms. Although unfinished, this work led directly to her summer 1844 discussions with Dr. William Benjamin O'Shaughnessy Greig about a mathematical model of the brain.

During this period, Ada visited the electrical experimenter Andrew Crosse at Fyne Court to learn laboratory techniques. His dramatic apparatus, voltaic piles, long copper wires, Leyden jars, and high-voltage discharges captivated her imagination and deepened her interest in electricity, nervous forces, and the possibility of describing mental processes mathematically. Newspapers called Crosse "The Thunder and Lightning Man." Around this time, Ada also formed a brief emotional attachment to his son, John. Afterwards, much of their correspondence was destroyed to protect her reputation.

By late 1844, Ada was seeking new intellectual direction and structured mentorship. After Sir David Brewster published a notice about the Analytical Engine in the *Edinburgh Review*, she wrote to him for guidance, noting that "some physiological topics" might suit her well. She continued to visit Charles socially, though their conversations often drifted toward lighter subjects. Charles, frustrated by the lack of recognition for his engines, even proposed building a tic-tac-toe playing machine to tour the country; Ada talked him out of it, though he continued refining the Analytical Engine until his death.

**SCIENCE SPOTLIGHT**

### Andrew Crosse's Experiments and the Mathematics of Mind

Andrew Crosse conducted ambitious electrical experiments at his home at Fyne Court, exploring how electricity shaped matter through controlled, repeatable setups. His work treated electrical phenomena as measurable, comparable, and analyzable patterns, an approach that resonated deeply with Ada.

For Ada, the importance of Crosse's experiments was not the spectacle but the method. His systematic way of probing natural forces strengthened her belief that one might understand complex processes, whether electrical, physiological, or mental, through symbolic relationships. This idea fed directly into her developing concept of a "calculus of the nervous system," a framework for expressing how thoughts and sensations might combine and transform in structured ways.

Crosse's laboratory offered Ada something rare: direct exposure to experimental science. It reinforced her conviction that patterns and operations could help study the mind, laying early groundwork for her attempts to describe mental processes mathematically.

## Dealing with Family Issues and Ill Health

Ada's life was far from easy, and she faced persistent family and marital challenges. Her marriage to William was often strained. Had he possessed a stronger personality or been better matched to her intellectual abilities, some conflicts might have been avoided. As their children grew older, Ada found the demands of motherhood increasingly stressful.

William pursued an endless series of elaborate construction projects, including towers and tunnels. To finance them, they frequently turned to Ada's mother, Annabella, who often made matters difficult and ultimately dominated the family. Ada also struggled with overindulgence in wine, which worsened when drinking gradually replaced meals. At one point, she even considered writing a scientific study on the effects of opium and wine, drawing from her own experiences.

Ada's social life added further complications. She kept company with young men in ways deemed scandalous at the time, flirting with several acquaintances. She may have had affairs, which led William to destroy over 100 of her letters to protect both Ada and the family.

By around 1850, Ada fell out with her mother, likely after discovering that Annabella had lied to her for years about her father, Lord Byron. Annabella had spent much of her life trying to shape Ada to be as unlike her father as possible. The revelation profoundly affected Ada, prompting her to explore her roots in a melancholic way. During a visit to her father's ancestral home in 1850, she wrote: "I seem to be in the mausoleum of my race. What is the good of living when thus all passes away and leaves cold stone behind … I feel as if I had become a stone monument myself. I am petrifying fast."

Ada had faced intermittent health problems since childhood. She struggled to recover from the birth of her second child and suffered what she described as heart or rheumatic attacks in the 1840s. Victorian medical practice prescribed laudanum (an opium tincture), morphine in small doses, and other sedatives for chronic pain and stress.

By late 1841, the pressures of early marriage, motherhood, a difficult third pregnancy, and her mother's constant supervision left Ada overwhelmed, exhausted, and intellectually frustrated. In early 1842, she suffered a nervous breakdown, becoming physically and emotionally unwell. Her mother interpreted this as a "nervous" or "moral" crisis, and Ada spent many months moving between doctors and various "cures," including mesmerism (hypnosis), all while observing the effects on what she called "that portion of the material forces of the world entitled the body of A.A.L."

The period kept Ada away from intellectual work, but by mid-1842, she gradually resumed her correspondence and studies. By late 1842, she was ready to engage in the intense mathematical work that would culminate in her 1843 collaboration with Charles.

In the mid-1840s, Ada frequently suffered migraines, chronic uterine pain, and anxiety, compounded by emotional turbulence and stress from accumulating gambling debts. Her doctor prescribed laudanum, which typically consisted of 10% opium dissolved in alcohol (usually wine or brandy). Over time, her use became regular and heavy, though still medically supervised. This led to dependence and periods of delusion. Despite these challenges, Ada remained optimistic in 1846, writing: "Nothing is needed but a year or two more of patience & cure."

Eventually, she overcame her addictions and delusions, but she became obsessed with betting on horse races, accumulating significant gambling debts during the 1840s.

## Ada's Gambling Habit

After the publication of her translations and Notes on the Analytical Engine, Ada became fascinated with the idea that probability theory could be used to predict horse race outcomes. She discussed these ideas with Charles, her tutor De Morgan, and other mathematically minded friends.

What began as an intellectual experiment with small bets soon escalated into enormous debts. Ada formed a secret syndicate with male friends, including John Croker and John Crosse, attempting to build a mathematical model for placing large, successful bets. The system failed catastrophically, leaving her owing thousands of pounds, equivalent to tens of thousands today, to the syndicate.

Ada tried to hide the debts from her husband, William, until it became impossible, leading to family conflict. Reports vary, but she may have wagered £20,000, or perhaps even £40,000 (roughly $7 million today) on horse races. She pawned her jewelry at least twice to pay off debts, highlighting the severity of her financial difficulties.

## Death

By the early 1850s, the Lovelaces lived primarily in London, but in 1851, Ada fell seriously ill. Initially, doctors advised her to spend time at the seaside. Eventually, they acknowledged she had cancer, likely uterine (endometrial) cancer, possibly involving the cervix, endometrium, or uterine body. She endured severe abdominal pain and hemorrhage, which were likely worsened by the now almost obsolete medical practice of bloodletting.

In mid-1851, Charles accompanied the frail Lovelace to the Great Exhibition, advising her to "put on worsted stockings, cork soles and every other thing which can keep you warm." However, he left disappointed because his machines were not on display there.

Ada continued her intellectual work and correspondence as much as her illness allowed. She started using cannabis when opium wasn't enough for her pain. By January 1852, her suffering had intensified, yet her mind remained sharp and vigorous. Her husband recorded: "Her mind was invigorated by the society of the intellectual men whom she entertained as guests. ... She mastered the mathematical side of a question in all its minuteness ... her power of generalisation was indeed most remarkable, coupled as it was with that of minute and intricate analysis."

Charles was a reliable intellectual partner, and their ongoing conversations built a foundation of genuine admiration and respect. By August 1852, Ada reflected on mortality: "I begin to understand Death; which is going on quietly & gradually every minute & will never be a thing of one particular moment."

On August 19, she invited Charles Dickens to visit and read an account of death from one of his books. Dickens later remarked on her strength and composure, despite her terminal illness.

In her last months, Ada underwent a religious transformation, possibly influenced by her mother, repenting for her previous conduct. Annabella moved into the house, taking control of medical decisions, correspondence, and visitors. She also became Ada's executor, and this role shaped how people would remember her daughter's life and legacy.

One of the last people outside the family to see Ada alive was Dickens, who read the death scene of young Paul Dombey from *Dombey and Son*. On September 1, 1852, Ada made a mysterious confession to her husband, which shocked him so deeply that he emotionally withdrew for the rest of her life. That same day, she wrote: "It is my earnest and dying request that all my friends who have letters from me will deliver them to my mother, Lady Noel Byron, after my death."

Ada succumbed to uterine cancer on November 27, 1852, at 36 years old, in St James's Square, Marylebone, London, the same age at which her father, Lord Byron, had died. She endured a slow and painful decline marked by wasting and organ failure.

Florence Nightingale, a friend and nursing pioneer, reflected on her resilience: "They said she could not possibly have lived so long, were it not for the tremendous vitality of the brain, that would not die."

At Ada's request, they buried her next to her father in the grounds of the Church of St. Mary Magdalene, Hucknall, England, near Lord Byron's ancestral home at Newstead Abbey. The funeral was small; neither her mother nor Charles attended. They placed her coffin side by side with her father's in the Byron family vault, which Sir John Byron had first built.

After her death, Annabella collected Ada's early poems, letters, moral reflections, and childhood writings, including the sonnet "The Rainbow." She used these selections to shape Ada's posthumous image as gentle, spiritual, and morally upright, a narrative intended to overshadow recent scandals.

Initially, Ada had asked Charles to serve as her executor, but ultimately her husband managed her estate. He oversaw legal negotiations with John Crosse, which resulted in the destruction of many letters, settled Ada's gambling debts, and handled the distribution of her personal effects and the guardianship of their children.

An obituary in *The London Examiner* summed up her character: "With an understanding thoroughly masculine in solidity, grasp and firmness, Lady Lovelace had all the delicacies of the most refined female character."

## Ada's Children

Ada's three children followed very different paths:

Byron King-Noel, her firstborn, was intense, restless, and frequently difficult, and was the child Ada most worried about. Moody, rebellious, and academically inconsistent, Byron often clashed with his father. Adventure and the military fascinated him, but beneath his bravado, he was deeply sensitive, and Ada feared he had inherited the "Byron temperament." At age 12, in 1848, he embarked on a journey that eventually took him to Tasmania, Chile, and California. He became a naval officer before Ada's death but later deserted. Byron died at 26 while working in a shipyard in England.

Anne Blunt, the middle child, most closely resembled Ada in temperament and intellect. A brilliant linguist (Arabic, French, German), artistically talented, and independent-minded, she displayed early traits of exploration, scholarship, and determination. Anne later married a somewhat wild poet, spent many years in the Middle East, and became Lady Anne Blunt, co-founder of the Crabbet Arabian Stud and a significant figure in Middle Eastern travel.

Ralph King-Milbanke, the youngest, was gentle, steady, and academically capable. Deeply attached to family, he spent most of his life on the family estate, eventually inheriting the Lovelace title and becoming the 2nd Earl. A desire for stability marked his teenage years, contrasting with Byron's turbulence and Anne's intensity. In 1905, a year before his death, Ralph published a memoir reflecting on the family, famously remarking: "Lord Byron's life contained nothing of any interest except what ought not to have been told."

Ada's husband, William, outlived her by 41 years and eventually remarried.

## Babbage's Machines and Ada's Contribution

Charles lived another eighteen years after Ada passed away in 1871. He attempted to resume work on the Analytical Engine in 1856 but made little progress. Although he never completed the Difference Engine, a Swedish company displayed part of it at the Great Exhibition. After his death, many documents and spare parts passed to his son, Major-General Henry Babbage, who published several papers and privately assembled a few devices, including part of the Analytical Engine's Mill. The surviving fragment of the Difference Engine eventually went to the Science Museum in London.

During his lifetime, Charles completed neither the Difference Engine nor the Analytical Engine, which remained an elaborate design rather than a working device. Years later, an extensive archive of his papers, found in a lawyer's family trunk, revealed many drafts of Analytical Engine expositions dating from the 1830s onward, with titles such as *Of the Analytical Engine* and *The Science of Number Reduced to Mechanism*. Why he never published them remains unclear.

While earlier thinkers like Pascal and Leibniz imagined automatic calculating machines, Charles's proposals marked a dramatic leap, containing all the essential features of a modern computer. Because they never built the Analytical Engine, Ada could not test her program. However, her idea of a machine repeating instructions, what we now call looping, remains a foundational principle of programming.

## Ada's Vision of Computing

Ada did not envision today's electronic computers. In her time, the term "computer" referred to a person, often a woman, who performed complex calculations in fields such as astronomy, ballistics, and engineering. She imagined a machine that could follow a predefined sequence of instructions to solve mathematical problems automatically.

Her work showed the potential of programmable machines and provided the first detailed explanation of how such a device could execute instructions.

More importantly, her vision extended far beyond arithmetic, anticipating many of the capabilities of modern computers.

The republication of Ada's Notes on Charles Babbage's Analytical Engine in 1953 allowed a new generation of computer scientists to recognize their extraordinary significance. Although she did not invent the machine itself, she possessed a remarkable ability to translate Babbage's abstract ideas into concepts that others could understand and to imagine possibilities that even he had not fully envisioned.

---

**SCIENCE SPOTLIGHT**

### What Is a Computer Program?

A computer program is a collection of instructions that tells a computer how to perform a task. Programmers arrange instructions in a specific order so that the machine knows exactly what actions to take and when to take them. Modern programs may contain millions of lines of code, but they all follow the same fundamental principle: a sequence of instructions designed to solve a problem or produce a result.

Many historians regard Ada Lovelace's Bernoulli number program as the first published computer program because it described, step by step, how the Analytical Engine could carry out a complex calculation automatically. Although no one built the machine during her lifetime, her work showed how a programmable device could execute a sequence of operations without human intervention, a concept that lies at the heart of modern computing.

---

## Challenges and Recognition

The formal barriers to women in science during the 1830s were severe. The Royal Society did not admit women, universities were largely inaccessible, and the scientific establishment tolerated women only as patrons, muses, or dinner guests, not as peers. Ada navigated these restrictions through her social position as an earl's daughter and countess, and through personal relationships with scientists willing to treat her seriously.

For decades, some modern commentators, especially male computer scientists, underestimated her contributions, viewing her as a publicist rather than a thinker. More recent research has confirmed that Ada's work was fundamental to the early development of computing.

## Alan Turing and Ada's Lasting Influence

Ada's influence continued long after her death. Nearly a century later, Alan Turing and other pioneers of modern computing revisited her ideas, particularly her argument that machines could follow instructions but could not originate ideas on their own. Turing took her arguments about machine intelligence seriously, addressing what he famously called "Lady Lovelace's Objection." Ada's insight that a machine could follow instructions yet not originate ideas remained central to debates on artificial intelligence, confirming her vision was far ahead of her time.

Turing drew inspiration from Ada's Notes, even though Charles's Analytical Engine never moved beyond the design stage. They addressed fundamental questions about what machines could and could not do. In her notes to Luigi Menabrea's paper, she wrote: "The Analytical Engine has no pretensions whatever to originate anything. It can do whatever we know how to order it to perform. It can follow analysis, but it has no power of anticipating any analytical revelations or truths. Its province is to assist us in making available what we are already acquainted with."

This remark, later dubbed "Lady Lovelace's Objection" by Turing, identified the limits of computational machines: a program can only execute what humans direct it to do. Ada anticipated that AI, even in principle, could not create truly original content without human input, an observation that remains relevant today, as all AI algorithms rely on human-generated data to operate.

Turing addressed Ada's objection in his 1950 paper, *Computing Machinery and Intelligence*, arguing that she had underestimated the potential of machines because she could not foresee the enormous speed, memory, and processing power of modern computers. Turing proposed that if a computer could mimic human responses convincingly enough, an interrogator could not reliably distinguish them from a human, and therefore, it could possess artificial intelligence. This became known as the Turing Test, in contrast to the "Lovelace Test," which evaluates whether a machine can display creativity independent of its programming.

Ada's work was further highlighted when, in 1953, British physicist Bertram Vivian Bowden republished her notes as an appendix to *Faster Than Thought: A Symposium on Digital Computing Machines* and reintroduced her work to a new generation of computer scientists.

Ada's writings were no longer seen as a historical curiosity but as evidence that she had conceptually understood computers and programming decades before they existed and laid the groundwork for thinking about computational universality. The Analytical Engine, with its arithmetic logic unit, conditional branching, loops, and integrated memory, meets the criteria for a computationally universal device. In modern terms, it was the blueprint for a general-purpose computer.

---

**SCIENCE SPOTLIGHT**

## What Is Artificial Intelligence?

Artificial intelligence (AI) refers to systems designed to perform tasks that typically require human intelligence, such as recognizing patterns, making predictions, or generating language. Modern AI relies on machine learning, in which algorithms learn from data rather than following fixed, handwritten instructions. These systems detect patterns, adjust their internal models, and improve with experience.

One branch of machine learning powers large language models (LLMs), systems trained on vast amounts of text to generate human-like responses, summarize information, translate languages, or assist with reasoning. They do not "think" like humans, but they excel at identifying statistical patterns and producing coherent, context-aware output.

Alan Turing engaged with Ada Lovelace's ideas because she raised one of the earliest questions about machine creativity: could a machine ever originate something new, or only do what we tell it to do? Turing argued that if a machine's responses were indistinguishable from a human's, we should regard it as demonstrating intelligence, directly addressing Ada's skepticism.

The debate Ada began continues today. Researchers still ask whether AI systems can truly create, or whether they simply remix patterns from their training data. As AI grows more capable, questions about originality and what remains uniquely human remain central in computer science, philosophy, and the arts.

## The Ada Programming Language

In the 1970s, the U.S. Department of Defense sought a high-level programming language to unify dozens of incompatible systems. They originally named the language DoD-1 before renaming it ADA in 1979 to honor Ada. On December 10, 1980, the military approved the language reference manual. The military standard MIL-STD-1815 reflects the year of Ada's birth.

ADA remains in use today, particularly in real-time systems across aviation, healthcare, transportation, finance, infrastructure, and space industries. Its first iteration, ADA 83, was procedural, while ADA 95 introduced object-oriented capabilities. Though no longer mandated for U.S. Department of Defense use, ADA continues to support the engineering of large, reliable programs.

## Reconstruction of the Analytical Engine

In 1991, under the leadership of Doron Swade, curator of computing at the London Science Museum, a team completed the full-size calculation element of Charles's Difference Engine. By 2002, they had built a full-scale working model, carefully using only materials and precision levels available in Charles's time. The 17-year project showed that Charles's Difference Engine could indeed have been built in the 19th century, fulfilling Ada's vision of programmable computation.

## Preservation of Her Work

Although Ada requested that her mother receive her letters before her death, thousands of her document pages remain scattered globally. Her youngest son later organized the letters returned to the family, but destroyed some of them.

The Bodleian Library in Oxford now holds the Lovelace family papers, and scholars globally can access them there. Thanks to the generosity of the Clay Mathematics Institute and Ada's descendants, a digital edition of her mathematical papers is now available online, ensuring that her pioneering work continues to inspire new generations.

## Legacy

The first edition of Ada's work, which includes what many consider the world's first computer algorithm, sold for £95,000 ($125,000) at auction in 2018. The sale highlighted the enduring historical and technical significance of her writings. On 25 July 2018, the U.S. Senate formally honored her life and legacy.

Other initiatives celebrating Ada's contributions include:

- The Ada Lovelace Project: A Rhineland-Palatinate competence center inspiring girls and young women to pursue mathematics, science, and technology careers.
- The Ada Lovelace Institute: Founded in 2018 in collaboration with the Alan Turing Institute, the Royal Society, and other partners, aiming to ensure that data and AI work for society.

- Ada Lovelace Day: Celebrated internationally each year on the second Tuesday of October since 2009, highlighting women's contributions to STEM and creating role models for future generations.
- The Ada Lovelace Excellence Scholarship: Established in 2026 by the University of Southampton, recognizing academically talented students in digital technology and promoting diversity in the field.

In Ada's time, people seldom recognized women as serious scientific thinkers. Her skills often seemed masculine to others. Charles, who famously called her the "Enchantress of Numbers," wrote: "She has thrown her magical spell around the most abstract of sciences and has grasped it with a force which few masculine intellects (in our own country at least) could have exerted over it."

Ada's intellectual vision, bridging mathematics, logic, and programmable machines, continues to inspire computing, artificial intelligence, and STEM education worldwide. Her insights remain foundational, proving that a 19th-century mind could foresee the potential of machines centuries before they became a reality.

## Why She Matters Today

Ada Lovelace's ideas continue to shape the digital world. Her insight that machines could manipulate symbols and numbers laid the conceptual foundation for modern computing, from programming languages to artificial intelligence.

Ada understood that a machine could follow a sequence of instructions to perform complex tasks, a principle that underlies every programming language used today. Her Notes on the Analytical Engine anticipated loops, subroutines, and algorithmic thinking long before electronic computers existed.

While others only saw a sophisticated calculating machine, Ada envisioned a device that could generate music, graphics, and patterns, as well as anything else expressible symbolically. This extraordinary leap of imagination foreshadowed modern creative computing, data visualization, and generative technologies.

Her work helped establish the idea that computation is not simply an engineering challenge but a new way of thinking. Computer science education, software design, and algorithmic research all trace part of their intellectual heritage to the questions she asked about how machines process information.

Every smartphone app, video game, search engine, and artificial intelligence system relies on the same fundamental principles that Ada recognized in the Analytical Engine. Although she never saw a working computer, her vision anticipated a world in which programmable machines would become essential tools for science, communication, business, and everyday life.

Ada's legacy continues to inspire generations of women and girls entering mathematics, engineering, and computer science. Her story, brilliant, bold, and far ahead of its time, reminds us that some of the greatest discoveries begin not with a machine, but with the imagination to see what one might become.

## Timeline

| Year | Event |
|---|---|
| 1815 | Born in London |
| 1816 | Parents divorced |
| 1824 | Lord Byron died |
| 1828 | Wrote Flyology |
| 1829 | Contracted measles |
| 1832 | Met William Turner |
| 1833 | Met Mary Sommerville and Charles Babbage; viewed a prototype of the Difference Engine |

| | |
|---|---|
| 1834 | Analytical Engine envisioned |
| 1835 | Married William King |
| 1836 | Byron King-Noel born |
| 1837 | Anne Isabelle born |
| 1839 | Ralph born |
| 1842 | Began work on the Translation; suffered a nervous breakdown |
| 1843 | Published the Translation and Notes (A–G), including the first published computer algorithm |
| 1840s | Developed a mathematical betting system for horse racing |
| 1844 | Focused on a mathematical model of the brain; met Andrew Crosse; Had relationship with John Crosse |
| 1852 | Died aged 36 |

"Imagination is the Discovering Faculty, pre-eminently. It is that which penetrates into the unseen worlds around us, the worlds of Science."

—*Ada Lovelace*

# ROSALIND FRANKLIN

(1920-1958)

## Chemist and X-ray Crystallographer

Her pioneering research provided crucial evidence for understanding the structure of DNA.

Original portrait illustration by Osmo, based on a historical reference photograph (c. unknown)

"The joy of science is such that no amount of publicity could equal the pleasure of discovering something yourself."

—*Rosalind Franklin*

Rosalind Franklin grew up in a family that valued service and tradition, yet she felt an unmistakable pull toward science. Guided by discipline and clarity of purpose, she pursued her studies despite the narrow opportunities available to women. Her precision and insight would illuminate the structure of life at the molecular level.

## Childhood

Rosalind Elsie Franklin was born on July 25, 1920, in Notting Hill, London, into an educated Anglo-Jewish family known for public service and philanthropy. Her father, Ellis Arthur Franklin, was a merchant banker at Keyser's Bank, and her mother, Muriel Frances Whaley Franklin, came from a family that also valued education and community responsibility. These principles shaped all five Franklin children.

Rosalind grew up between her older brother David and younger brothers Colin and Roland; her sister Jennifer was born when she was nine. Surrounded by boys, she often joined their activities and developed interests that differed from traditional expectations for girls. She preferred building things, drawing, photography, and reading over dolls, and she enjoyed hiking and sports, interests that stayed with her throughout life.

Her intelligence, especially in mathematics and science, was clear early on. At six, her aunt Mamie described her as "alarmingly clever," noting she did arithmetic for fun and "invariably gets her sums right." Ellis encouraged her curiosity and introduced her to scientific ideas. When she first learned how photographic development worked, she exclaimed, "It makes me feel all squidgy inside."

Rosalind was observant, perceptive, and sensitive. Though intellectually confident, she withdrew deeply when hurt or treated unfairly. Her mother said she would "figuratively curl up, like touching the fronds of a sea anemone." To those close to her, she was funny, mischievous, and sharp-witted, though her dislikes, as well as her loyalties, were enduring.

She approached everything with discipline and efficiency and struggled to understand why others did not. She could be blunt and intensely critical, especially in scientific matters, and had little patience for carelessness. Family discussions were lively debates on politics, science, and current events, and Rosalind excelled through logic and articulate argument, despite her quiet voice. Her independent nature often led her to challenge her father's views.

Ellis taught evening classes in electricity, magnetism, and First World War history at the Working Men's College. He had once hoped to become a scientist himself before the war interrupted his education. As she grew older, Rosalind helped him with campaigns supporting the college, which provided educational opportunities to working-class men.

The Franklins also assisted Jewish refugees fleeing Nazi persecution. Rosalind even shared her room with a young refugee whose father the Nazis had imprisoned. The Franklin women were encouraged to apply their talents to charitable work, but Rosalind always aspired to be a scientist from a young age.

## Education

At age six, Rosalind attended Norland Place School in West London with her brother, Roland. In 1929, at the age of nine, she enrolled at Lindores School for Young Ladies in Bexhill-on-Sea. Her family chose this school partly because of the perceived health benefits of its seaside location for her fragile health.

In 1931, at eleven, Rosalind enrolled at St Paul's Girls' School in Hammersmith. 'Paulinas' followed a rigorous academic and athletic program designed to prepare them for demanding professions. St Paul's was one of the few girls' schools offering advanced chemistry and physics, and its headmistress, Miss Ethel Strudwick, encouraged independence and meaningful work.

Rosalind thrived academically. She excelled in science, mathematics, and languages, becoming fluent in French, proficient in Italian, and capable in German. Music was her weakest subject; composer Gustav Holst, the school's music director, once remarked that she had improved to singing "almost in tune."

Outside the classroom, she enjoyed cricket, hiking, and trekking. Although she often felt anxious before exams, she consistently earned high marks and academic prizes. During her years at St. Paul's, she became especially drawn to physics and chemistry, which would shape her future career.

By fifteen, Rosalind had already decided to become a scientist. Her mother later said that "all her life Rosalind knew exactly where she was going." Her father initially discouraged this ambition, given the limited opportunities for women in science, but Rosalind remained determined. Though her parents hoped she might eventually marry, they gradually supported her goals, even as some relatives feared a scientific career might make her "unmarriageable."

Rosalind graduated from St Paul's in 1938 and passed the entrance exam for the University of Cambridge. She joined Newnham College, one of Cambridge's women's colleges. Her acceptance caused tension at home when her father, opposed to university education for women, refused to pay her tuition. With strong support from her aunt and her mother, Ellis eventually relented and agreed to fund her studies.

## Newnham College, Cambridge

At eighteen, Rosalind began studying at Newnham College, one of the two Cambridge colleges that admitted women. Although women were not full university members until 1948, Newnham offered rigorous preparation for Cambridge degrees. She chose physical chemistry, a discipline grounded in physics, thermodynamics, quantum theory, kinetics, and mathematics, which suited her analytical mind.

Throughout her studies, she wrote weekly letters home. In the summer of 1940, she told her father: "Science and everyday life cannot and should not be separated... it is based on fact, experience and experiment." During university vacations, she volunteered at Woburn House with the German Jewish Refugee Committee, reinforcing her growing sense of social responsibility.

A second-year course in crystal physics introduced her to crystallography and the foundational work of William Henry Bragg and William Lawrence

Bragg. Though she never studied directly under them, the course exposed her to symmetry, space groups, and crystal structures, concepts that later shaped her research.

As World War II began, Rosalind described nights interrupted by air raid sirens, sleeping in college shelters, and serving as a fire watch warden. Despite the upheaval and the departure of many researchers for war work, she excelled academically and placed second in her year. Her instructor, Frederick Dainton, later remembered her as direct, rigorous, and honest, someone who believed "the facts will speak for themselves."

For her final undergraduate project, Rosalind studied the polymerization of formic acid and acetaldehyde, exploring how simple molecules form larger chains. This work brought her into close contact with her supervisor, Professor Ronald George Wreyford Norrish. Norrish, a leading figure in Cambridge physical chemistry and a later Nobel laureate, had a reputation for being demanding and dismissive, especially toward junior researchers and women.

Their relationship deteriorated when Rosalind discovered that one of the assumptions behind his calculations was incorrect. Her results consistently contradicted his predictions, and after repeatedly checking her work, she concluded the error was his. Confident that debate strengthened science, she challenged him. Norrish reacted defensively and insisted she repeat the experiments. Rosalind later wrote that he "became most offensive" when she stood up to him. After this, their relationship never recovered. He later criticized her interest in "raising the status of her sex to equality with men," reflecting attitudes common in science at the time.

In 1941, Rosalind completed the Natural Sciences Tripos with Second Class Honors in physical chemistry. Because Cambridge did not yet grant women full degrees, she received only a certificate acknowledging her studies. Although Newnham awarded her a research scholarship, she chose not to remain in academia during wartime and instead joined the British Coal Utilisation Research Association (BCURA), contributing directly to Britain's scientific efforts.

## The Study of Coal

BCURA was located on the Coombe Springs Estate near Kingston upon Thames and was an independent research organization supported by industry and the British government. During World War II, it focused on improving the efficiency and industrial use of coal, a resource vital to Britain's homes, factories, and wartime production.

Rosalind joined BCURA in 1942 as an Assistant Research Officer under John G. Bennett. She investigated the internal structure of coal, especially its porosity, the microscopic network of pores that determines how gases and liquids move through the material. Because coal's performance depended heavily on its internal architecture and behavior under heat, this work was essential.

Over the next four years, she studied how coal changed during carbonization, the process of heating coal without oxygen to produce coke. Coke, a nearly pure form of carbon, burned hotter and cleaner than raw coal and was strong enough to withstand the extreme conditions inside blast furnaces, making it indispensable for steelmaking.

Rosalind employed physical chemistry methods to determine the density, surface area, porosity, thermal decomposition, and gas absorption of various coals. Her goal was to comprehend how heat treatment affected their microscopic structure and the reasons behind their varied performance in industrial processes.

Through this work, she made a major discovery: coal contained extremely narrow molecular-scale constrictions that controlled how gases passed through the material, and the degree of constriction varied with carbon content. Heating changed the size and shape of these pathways, allowing some molecules to pass while blocking others. She described this behavior as a "molecular sieve" and became the first scientist to identify and measure these differences systematically, enabling coals to be classified by internal structure rather than appearance or chemistry alone.

Her findings significantly improved the understanding of coal behavior during carbonization and helped predict performance more accurately during wartime production. This research established Rosalind's reputation as an exceptionally careful and insightful experimental scientist.

---

**SCIENCE SPOTLIGHT**

### Carbonization

Carbonization is the process of heating coal in the absence of air to drive off water, gases, and tar, leaving behind coke, a nearly pure, porous carbon solid.

Heating coal causes its internal structure to reorganize. Volatile compounds escape, pores open, and carbon layers begin to align, changes that determine how strong, reactive, and heat-resistant the final coke will be, properties essential for steelmaking. As temperatures rise, coal's complex organic molecules break apart, releasing gases and allowing carbon atoms to rearrange into more ordered layers. This transition from a disordered organic solid to a partially graphitized carbon network controls porosity, strength, and reactivity. By measuring density, porosity, and gas absorption, researchers can track how coal transforms into coke and predict its industrial performance.

---

## Rosalind's Wartime Research and Service

Rosalind's research at BCURA had important practical applications during World War II. Fearing the use of chemical weapons, as in World War I, the British government issued gas masks to every civilian. Activated charcoal filters had already proven highly effective in protecting soldiers from poisonous gases.

Her studies of porosity and gas adsorption, measuring pore size distribution, surface area, and gas absorption, helped scientists understand how carbon materials trap and filter gases. She showed how coal's microscopic structure influenced its ability to absorb toxic substances. By identifying how heat treatment altered internal pathways, she helped determine which forms of carbon were most effective for filtration. These findings contributed to improvements in activated carbon and guided the production of more efficient gas mask filters during the war.

Alongside her scientific work, Rosalind also served as a London air raid warden during the Blitz, when British cities endured repeated German bombing raids. Wardens enforced blackout regulations, guided civilians to shelters, reported bomb damage, administered first aid, and assisted emergency services during attacks. Often among the first to arrive after explosions, they helped injured residents, maintained order, and offered reassurance during chaotic and dangerous nights. The work was physically demanding and sometimes perilous, especially during heavy raids.

## PhD and Carbon Research

Although Rosalind completed her studies at Cambridge in 1941, the university did not award degrees to women until 1948, when she finally received her Bachelor of Arts. Meanwhile, her research at BCURA formed the basis of her doctoral work, and in 1945 she earned a PhD for her studies on the physical chemistry of coal. Her thesis, *The Physical Chemistry of Solid Organic Colloids with Special Reference to Coal and Related Materials*, expanded her pioneering investigations into carbon structure.

Rosalind showed that different carbons behave very differently when heated. Some gradually reorganize into the orderly, layered structure of graphite; these graphitizing carbons loosen internally, allowing carbon sheets to grow and align. Others resist this transformation entirely. Their layers are tangled, cross-linked, or riddled with defects that lock the structure in place. These non-graphitizing carbons remain hard, porous, and disordered no matter how high the temperature. This distinction became one of Rosalind's most important early insights and remains central to modern carbon science.

In 1946, she published her first peer-reviewed paper, *Thermal Expansion of Coals and Carbonised Coals*, co-authored with Douglas Heath Bangham, further developing her "molecular sieve" hypothesis to explain variations in porosity among coals and carbonized materials.

At the BCURA, Rosalind enjoyed an unusual amount of independence for a junior researcher, especially for a woman in the 1940s. The environment suited her disciplined, methodical approach, and by the time she left in 1947, she had published five significant papers and established herself as one of Britain's most promising young physical chemists. Her growing reputation soon opened opportunities abroad.

## Life in Paris

During her final years at Newnham College, Rosalind became acquainted with the French scientist Adrienne Weill, who later became both a close friend and an important professional connection. After World War II, Rosalind sought opportunities beyond coal research. Though she had built a strong reputation at BCURA, she wanted to expand her skills into more advanced structural work. Writing to Adrienne, she joked she was "a physical chemist who knows very little physical chemistry, but quite a lot about the holes in coal."

At a 1946 carbon research conference in London, Adrienne introduced her to Marcel Mathieu, a senior figure in the French scientific system. Impressed by Rosalind's accomplishments, he helped arrange a position for her with Jacques Méring at the Laboratoire Central des Services Chimiques de l'État in Paris.

In 1947, Rosalind moved to Paris to begin her postdoctoral research, a period that became one of the happiest of her life. Having studied French at St Paul's, she arrived already comfortable with the language and soon became fully fluent. She embraced French culture, appreciating the food, intellectual atmosphere, and the more egalitarian scientific environment. She later remarked that in France, "women engaged as equals," a sharp contrast to many of her British experiences.

Her modest salary required careful budgeting, but she valued supporting herself. For several years, she lived in a small, rented room on Rue Garancière in the Left Bank, near the Sorbonne and the Luxembourg Gardens. Though simple and governed by strict rules, the location placed her at the center of Parisian cultural life. Rosalind quickly adapted to postwar Paris. She enjoyed local markets, French cooking, and evenings at lectures, concerts, films, and exhibitions. She also developed an appreciation for French fashion, adopting styles inspired by Christian Dior.

Paris deepened her love of travel and the outdoors. She spent weekends cycling through the countryside, hiking in the Alps, or traveling across Europe with friends, often choosing inexpensive third-class travel to meet people and practice languages. Friends remembered her as warm, lively, humorous, and sociable. Free from the tensions of British academic culture, Rosalind flourished both personally and professionally.

## Research at the Laboratoire Central des Services Chimiques de l'État.

On February 14, 1947, Rosalind joined Jacques Méring's laboratory at the Laboratoire Central des Services Chimiques de l'État on the Quai Henri IV in Paris. Méring was a pioneering Jewish French physicist whose work transformed the study of disordered and partially ordered materials. William Henry Bragg trained him at the Royal Institution, and he led one of France's key research groups in X-ray crystallography.

The laboratory environment differed sharply from the formal, hierarchical atmosphere Rosalind had known in Britain. Scientists from many countries worked collaboratively, and men and women operated as equals. Rosalind thrived in this setting, immersing herself in experimental research. Méring focused on materials that resisted traditional crystallography, substances with irregular or partly ordered structures such as rayon, coal, and complex carbons. His lab was one of the few equipped to analyze such challenging samples, offering Rosalind both freedom and technical sophistication.

Under Méring's guidance, she mastered advanced X-ray diffraction and crystallographic methods, learning to interpret patterns from poorly ordered materials and apply mathematical analysis to reveal internal structure. Her precision and patience made her exceptionally suited to this demanding work.

During her Paris years, Rosalind also deepened her engagement with William Lawrence Bragg's research. Her notebooks show extensive use of Bragg's Law and related techniques as she investigated coal structure and the transformations that occur as carbon approaches graphite. Her work helped lead the way in using X-rays to investigate disordered substances, highlighting important differences between graphitizing and non-graphitizing carbons and improving knowledge of carbon microstructure.

**SCIENCE SPOTLIGHT**

## X-ray Crystallography

X-ray crystallography is a technique used to determine the atomic structure of a crystal by measuring how X-rays scatter when they strike its regularly arranged atoms. The atoms act like a three-dimensional diffraction grating, producing a pattern of spots that encodes the crystal's internal structure.

In Rosalind's time, she recorded these patterns on photographic film, while today, digital detectors and synchrotron sources capture them with far greater precision. Each spot reflects X-rays from a specific set of atomic planes. Its position reveals the angle of diffraction, while its intensity shows how strongly that plane scattered the beam. Together, these measurements allow scientists to reconstruct a three-dimensional map of atomic positions.

A single image is never enough. Rotating the crystal through many small angles produces hundreds or even thousands of diffraction patterns. The system combines these patterns into a complete three-dimensional model. This method reveals structures far too small for any microscope and has been essential for understanding minerals, metals, polymers, proteins, and other biological molecules.

**SCIENCE SPOTLIGHT**

## X-ray Diffraction vs. X-ray Crystallography

All crystallography relies on X-ray diffraction, but not all diffraction studies are crystallography. X-ray diffraction (XRD) is the broader technique used to study crystals, powders, fibers, polymers, and partially ordered materials. X-ray crystallography is the specific application of XRD to well-ordered crystals to determine their three-dimensional atomic structure. It requires long-range order and sharp diffraction patterns, and is the method used to solve the structures of DNA, proteins, minerals, and many small molecules.

The work was labor intensive: every diffraction image had to be developed, measured, and analyzed by hand. Rosalind's meticulous care produced images of exceptional quality. Colleague Vittorio Luzzati praised her "golden hands." In 1948 alone, she published five papers on coal and carbon; her first *Nature* letter appeared in 1950, followed by major work in the *Proceedings of the Royal Society*. Her research advanced carbon science and contributed to the development of heat-resistant carbons and carbon fibers. Méring regarded her as one of his finest students.

The Paris laboratory's collaborative, emotionally expressive atmosphere also shaped her. Physicist Geoffrey Brown joked it was like "a traveling opera company," with arguments that quickly dissolved into friendship. Rosalind earned deep respect for her honesty, rigor, and integrity.

By 1950, she had started to consider returning to Britain. Family expectations, new opportunities, and conversations with Dorothy Hodgkin and theoretical chemist Charles Coulson, who encouraged her to apply her diffraction expertise to biological molecules, helped guide her next steps.

# The DNA Project Begins

## King's College London and the DNA Project

In January 1951, after four productive years in Paris, Rosalind accepted a three-year Turner & Newall Fellowship to join the Medical Research Council's Biophysics Unit at King's College London. Biophysicist John Randall directed the unit, with physicist Maurice Wilkins as deputy director. Randall recruited Rosalind for her growing international reputation in X-ray diffraction and her exceptional skill in interpreting complex diffraction data.

Rosalind initially expected to study proteins and lipids in solution, examining how their structures changed when heated or dehydrated. Although highly accomplished in physical chemistry and crystallography, she had little background in biology and admitted she was "most ignorant about all things biological," adding that many X-ray crystallographers likely began the same way.

Shortly before she arrived, however, Randall reassigned her to work on DNA fibers. Although scientists identified DNA in 1869, its structure and role in cells remained unknown, making it one of the most important scientific problems of the era. The race to solve it involved several major laboratories: Linus Pauling's group in the United States, the team at King's, and the researchers at the Cavendish Laboratory in Cambridge, where James Watson and Francis Crick had begun their own efforts.

The DNA samples at King's were unusually pure fibers obtained by Wilkins from Swiss chemist Rudolf Signer in 1950. With these, Wilkins and graduate student Raymond Gosling had already produced promising early X-ray photographs and identified two humidity-dependent forms of DNA, which Wilkins referred to as the A-form and B-form.

Although Wilkins had started the DNA work at King's, Randall intended Rosalind to take primary responsibility for the X-ray diffraction studies. In a December 4, 1950, appointment letter, he informed her that she and Gosling would lead the experimental effort. However, Randall failed to explain this clearly to Wilkins. When Rosalind arrived, Wilkins was away on vacation and unaware that she now had independent authority over the project. He assumed she was joining to assist him, not to direct the work herself.

The misunderstanding created tension almost immediately. Rosalind saw herself as an independent scientist with her own research program, while Wilkins believed the project remained under his leadership. This miscommunication quickly hardened into resentment and set the stage for a difficult working relationship that would later shape the course of DNA research.

## Challenges at King's College

The atmosphere at King's College differed sharply from the collaborative environment Rosalind had enjoyed in Paris. Although Randall was relatively supportive of women scientists, women still faced significant institutional barriers within British scientific culture during the 1950s.

Rosalind received an excellent laboratory, high-quality DNA samples, and strong research funding. However, women remained excluded from several important social and professional spaces at King's, including the men-only dining and common rooms where many informal scientific discussions took place. These exclusions often isolated female researchers from important conversations and collaborations. The situation frustrated Rosalind deeply because it contrasted so sharply with the more egalitarian scientific culture she had experienced in France, where men and women worked and socialized more equally.

At lunchtime, Rosalind often joined mixed groups of MRC researchers who found her intelligent, witty, and engaging. While outside the laboratory, she could be lively and sociable. However, once she returned to her experiments, her intense concentration and direct manner frequently made her appear stern and uncompromising.

Several colleagues viewed Rosalind as "too French" in both style and temperament. She often skipped tea breaks to continue working and approached scientific disagreements with the same direct intellectual debate she had embraced in Paris. Some colleagues found her assertiveness difficult, particularly within the more reserved and hierarchical culture of King's. Her friend Norma Sutherland later recalled that Rosalind's manner could sometimes seem brusque or confrontational and that she appeared unaware of the hostility this occasionally created.

Gosling later contrasted Rosalind's forceful, logic-driven debating style with Wilkins's quieter personality, noting that when she believed someone was wrong, she challenged them directly and expected a reasoned defense. Wilkins, by contrast, avoided confrontation.

## Rosalind Franklin and Maurice Wilkins

Rosalind and Wilkins had very different personalities and working styles. Wilkins was introverted, cautious, and uncomfortable with open conflict. Rosalind was intellectually assertive, highly analytical, multilingual, and accustomed to vigorous scientific debate.

Her habit of maintaining intense eye contact, speaking directly, and challenging weak arguments unsettled some colleagues. Wilkins, in particular, often found her manner difficult. Rosalind, meanwhile, became increasingly frustrated by what she saw as poor communication and unclear authority within the laboratory.

Wilkins later admitted that the situation had become "a bad bargain" for him after he agreed to transfer the valuable Signer DNA samples to Rosalind's control. While Rosalind worked with the exceptionally pure DNA fibers from Signer, Wilkins continued working with lower-quality samples from biochemist Erwin Chargaff.

At King's, Rosalind was technically the more experienced X-ray diffraction specialist. Wilkins himself acknowledged this difference, later remarking, "We were amateurs, but she was a professional."

Misunderstandings about status and authority intensified the conflict between them. Rosalind considered herself Wilkins's equal, but many colleagues assumed she was subordinate to him because she was both younger and female. Wilkins also referred to her as 'Rosy, ' a nickname Rosalind disliked and never used herself. As tensions increased, Rosalind became increasingly reluctant to share preliminary findings openly with Wilkins. Communication between the two deteriorated into mutual distrust and resentment, creating an atmosphere that hindered collaboration.

Scientifically, the two researchers also pursued different approaches. Rosalind concentrated much of her effort on the more detailed but technically difficult A-form of DNA, while Wilkins focused primarily on the B-form, whose early diffraction patterns appeared more obviously helical.

---

**HISTORICAL SPOTLIGHT**

## Maurice Wilkins

Maurice Wilkins (1916–2004) was born in Pongaroa, New Zealand, and moved to England as a child. He studied physics at St John's College, Cambridge, and completed a PhD at Birmingham University on how solids absorb and release light and heat. During World War II, he helped improve radar screens and later worked on uranium isotope separation. His time with the Manhattan Project in 1944 left him uneasy after the atomic bombings.

Disillusioned with weapons research, Wilkins, like John Randall, shifted toward interdisciplinary biophysics. They established new laboratories at St Andrews and later at King's College London, where Wilkins applied physical methods to biological problems. At King's, he improved reflecting microscopes for visualizing nucleic acids and studied base-pair orientation in Tobacco Mosaic Virus, gradually focusing on nucleic acid structure.

After receiving high-quality Signer DNA in 1950, Wilkins and Raymond Gosling began systematic X-ray diffraction studies. Wilkins remained at King's for the rest of his career, applying X-ray methods to nerve membranes, RNA, and other biological structures. As biophysics matured into a distinct discipline, he became one of its central institutional figures.

---

**SCIENCE SPOTLIGHT**

### What DNA Is

DNA, or deoxyribonucleic acid, is the long, spiraling molecule that carries the genetic instructions for building and maintaining every living organism. It acts as the body's master blueprint, storing information in a precise chemical code.

A DNA molecule comprises two strands twisted into a double helix. Each strand comprises repeating units called nucleotides. These nucleotides include a sugar, a phosphate group, and one of four bases: adenine (A), thymine (T), cytosine (C), or guanine (G). The bases pair in a strict pattern: A with T, C with G, forming the "rungs" of the helix.

The sequence of these bases encodes instructions for making proteins, guiding everything from cell growth to inherited traits. Cells copy DNA with remarkable accuracy when they divide. This accuracy ensures the passage of genetic information from one generation to the next.

DNA explains heredity, reveals evolutionary relationships, and underpins modern medicine, from genetic testing to targeted therapies. The discovery of its structure opened the door to nearly every major advance in molecular biology.

## Rosalind Franklin's DNA Research

At King's, Rosalind rebuilt and modernized the X-ray diffraction laboratory, believing Randall had placed her in charge of the DNA project. Working closely with graduate student Raymond Gosling, she focused on determining the structure of deoxyribonucleic acid.

Before her arrival, scientists already knew that DNA had a repeating structure. William Astbury and Florence Bell used X-ray diffraction in the late 1930s, revealing a repeating structure of its components every 27 angstroms and the stacked bases spaced about 3.4 angstroms apart, although they did not identify a helix. Physicist Alexander Stokes had mathematically shown that DNA was likely helical, but the number of strands remained unknown.

Despite the strained atmosphere at King's, Rosalind refined her experimental methods and steadily improved the quality of her diffraction images. She prepared finer DNA fibers, developed more sophisticated techniques, and worked largely in professional isolation, with Gosling as her primary collaborator.

### Rebuilding the Experimental System

Rosalind and Gosling spent months redesigning and assembling specialized equipment to increase experimental accuracy. Using a fine-focus X-ray tube developed by Werner Ehrenberg and Walter Spear, Rosalind adjusted the system to produce an extremely narrow beam. She also designed an advanced humidity control chamber that allowed precise regulation of moisture around the DNA fibers. Gosling later recalled that during hydration experiments, the fibers sometimes expanded so dramatically they nearly "leaped off the stage." The hydrogen gas system was difficult and occasionally hazardous, and Gosling joked the leaks made him fear they might blow up part of King's.

Rosalind also perfected methods for extracting exceptionally fine DNA fibers and aligning them into ordered bundles. These improvements produced diffraction photographs far more sharply than any previously obtained. Her work demanded immense patience: she slowly hydrated and dehydrated fibers, monitored structural changes, and repeatedly adjusted conditions to get the clearest possible patterns.

## Unraveling DNA

### The A- and B-Forms of DNA

Drawing on her expertise in physical chemistry, Rosalind recognized that water content dramatically altered the structure of DNA fibers. Using highly purified Signer samples, she and Gosling produced far more precise diffraction photographs of the two known DNA forms. When relatively dry, the fibers became shorter, broader, and more crystalline (the A-form). When hydrated, they lengthened, thinned, and became less ordered (the B-form). Although Wilkins and Gosling had identified these forms earlier, Rosalind provided the first rigorous structural analysis of both.

She showed that the A- and B-forms could reversibly transform simply by adjusting the humidity. This insight proved crucial: earlier researchers had unknowingly examined mixtures of both forms, producing blurred, confusing diffraction patterns.

The A-form contained about 20% water and formed best near 75% relative humidity. The more hydrated B-form appeared at around 90% humidity and represented the DNA structure most adopted in living cells. Rosalind recognized that the additional water in the B-form allowed the molecule to adopt a more stable helix, with the bases aligned perpendicular to the fiber axis. By late 1951, most scientists at King's accepted that the B-form was helical. Rosalind initially believed both forms were helical, but after obtaining more detailed A-form data in 1952, she became less certain and insisted on further analysis before drawing firm conclusions.

Meanwhile, tensions in the laboratory escalated. In the summer of 1951, Wilkins presented some of Rosalind's findings at an international conference to support the helical model. Rosalind felt he had used his seniority to appropriate her results without proper acknowledgment and complained to Randall.

Randall responded by formally dividing the DNA project. He assigned Rosalind and Gosling, the more difficult A-form, using the best equipment and Signer fibers. Wilkins focused on the B-form, using lower-quality samples.

**HISTORICAL SPOTLIGHT**

### Raymond Gosling

Raymond Gosling (1926–2015) was born in London and studied physics at University College London. In 1949, he joined John Randall's biophysics group at King's College London as a graduate student, initially working under Maurice Wilkins, where he helped prepare highly ordered DNA fibers by carefully controlling humidity, an essential step in producing clear diffraction patterns. Under the supervision of Rosalind Franklin, he refined DNA sample preparation, and together they generated the famous Photo 51. Gosling completed his PhD in 1954 and moved into medical physics, spending much of his career at Guy's Hospital Medical School in radiological research and teaching.

## Watson, Crick, and the Race to Solve DNA

While Rosalind began her research at King's, another important effort to solve the structure of DNA was underway at the Cavendish Laboratory in Cambridge, England, under the direction of William Lawrence Bragg.

James Watson and Francis Crick were among the researchers at the Cavendish Laboratory. Unlike Rosalind, whose work focused on obtaining precise experimental evidence through X-ray diffraction, Watson and Crick approached the problem primarily through theoretical model building. Using cardboard cutouts, wire models, published chemical information, and data generated by researchers at King's, they attempted to construct a physical model of the DNA molecule.

During this time, American chemist Linus Pauling, working at the California Institute of Technology and already famous for solving the structure of proteins, was also racing to determine DNA's structure. The competition between the British and American groups intensified during the early 1950s.

**HISTORICAL SPOTLIGHT**

## James D. Watson and Francis H.C. Crick

James D. Watson, born in Chicago in 1928, studied zoology at the University of Chicago and completed his PhD at Indiana University on X-ray effects in bacteriophages. After hearing Maurice Wilkins present DNA diffraction images in 1951, he shifted to nucleic acid structure and joined the Cavendish Laboratory that fall.

Francis H. C. Crick (1916–2004), born in Northampton, trained in physics at University College London and worked during World War II on naval mines. After the war, he retrained in biology at the Strangeways Laboratory and joined the MRC Unit at the Cavendish in 1949. When Watson arrived in 1951, their complementary strengths, Crick's theoretical and structural insight and Watson's biological focus, formed a partnership aimed at solving DNA's structure.

Their collaboration produced the double-helix model in 1953 and helped launch molecular biology. Watson later joined Harvard in 1956, studying RNA and protein synthesis, and in 1968 became director of Cold Spring Harbor Laboratory, building it into a major genetics center. From 1988 to 1992, he served as the first director of the National Center for Human Genome Research, starting the Human Genome Project.

Crick remained at the MRC Laboratory of Molecular Biology, helping articulate the central dogma and contributing to deciphering the genetic code. In the 1970s and 1980s, he turned to theoretical neuroscience, exploring the biological basis of consciousness.

## 1951 DNA Lecture and the First Incorrect DNA Model

In November 1951, Rosalind delivered an important seminar at King's presenting her latest DNA research. Among the attendees was James Watson from the Cavendish Laboratory. She outlined key findings on both the A- and B-forms of DNA, explaining that the molecule existed in multiple states depending on humidity. Her notes described DNA as a "big helix with several chains," with the phosphate groups forming the exterior backbone. While the B-form appeared clearly helical, she remained cautious about drawing final conclusions for the more complex A-form, reflecting her commitment to strong experimental proof over speculation.

After returning to Cambridge, Watson and Crick quickly attempted to build their first DNA model. Their structure proposed a triple helix with the phosphate groups packed inside and the bases pointing outward. When Rosalind and Wilkins visited Cambridge to examine it, Rosalind immediately identified major flaws. She noted that packing the negatively charged phosphate groups together in the molecule's core was impossible and questioned where the surrounding water molecules would fit, exposing another structural impossibility. She concluded bluntly that the model did not match the experimental evidence and that Cambridge should leave DNA structure determination to King's. Her critique was correct: the model collapsed under scrutiny, and following complaints from King's, William Lawrence Bragg temporarily ordered Watson and Crick to stop working on DNA.

The pause ended in early 1953 when Linus Pauling published his own, incorrect triple-stranded model. Concerned Pauling might still solve the problem first, Bragg allowed Watson and Crick to resume their efforts. This time, they had access to far more accurate structural information, including key insights derived from Rosalind's diffraction data and analyzes.

## Growing Tensions

The failed Cambridge model widened tensions between the researchers at King's and Cambridge. Rosalind's direct criticism of Watson and Crick, combined with the increasingly strained atmosphere at King's, deepened divisions between Rosalind and Wilkins.

Wilkins, who had become friends with Crick, increasingly spent time at the Cavendish Laboratory. Meanwhile, some colleagues at King's began referring to Rosalind behind her back as the "Dark Lady," reflecting both the gendered attitudes of the time and the growing social isolation she experienced within the laboratory.

Rosalind could be uncompromising when she believed someone violated scientific standards or professional fairness. In one incident, she angrily confronted a conference acquaintance who planned to publish one of her ideas without proper acknowledgment, demanding co-authorship in a sharply worded letter. Her insistence on scientific rigor and proper credit sometimes made her appear intimidating to colleagues, particularly in the male-dominated scientific culture of the 1950s.

## Rosalind's Contribution

Rosalind's research had already established several critical facts about DNA. She showed that the molecule existed in two distinct forms, known as the A- and B-forms, depending on humidity, and that the B-form was strongly consistent with a helical structure. Her meticulous experiments also revealed that the phosphate backbone lay on the outside of the molecule and that DNA possessed regular repeating dimensions visible through X-ray diffraction. By carefully preparing highly ordered, hydrated DNA fibers, she produced diffraction images of unprecedented clarity and provided the precise measurements needed for accurate model building.

Rosalind's approach differed fundamentally from that of Watson and Crick. Rather than proposing bold theories based on incomplete evidence, she believed that rigorous experimental verification should support every conclusion. Having seen earlier structural models fail, she preferred to let the data speak for themselves. Although this cautious approach slowed her willingness to publish sweeping claims, it ensured that her results were exceptionally reliable.

Working with her graduate student, Raymond Gosling, Rosalind spent months refining every aspect of the experiment. She improved the orientation of the DNA fibers, upgraded the X-ray camera system, and developed precise humidity controls that allowed the fibers to be maintained in their hydrated B-form. They drew the highly purified Signer DNA samples into extremely fine, parallel bundles and suspended them within a sealed chamber. There, they exposed the samples to a finely focused X-ray beam for approximately sixty hours at around 92% humidity.

## Photo 51

On May 6, 1952, Gosling, working under Rosalind's direction, produced the famous diffraction image later known as "Photo 51," one of the clearest X-ray photographs ever taken of the B-form of DNA. Earlier experiments in the 1930s had unknowingly used mixed A-and B-forms of DNA, producing blurred and confusing patterns that were impossible to interpret clearly.

Photo 51 displayed a striking X-shaped diffraction pattern, a hallmark of a helical molecular structure, and was exceptionally sharp and detailed, far clearer than any previous image of hydrated DNA. Its remarkable clarity resulted from Rosalind's technical skill and painstaking attention to detail. Scientist John Desmond Bernal later described Photo 51 as "among the most beautiful X-ray photographs of any substance ever taken."

Photo 51 contained an extraordinary amount of structural information. Its prominent X-shaped diffraction pattern revealed that the molecule was helical, while the spacing and intensity of the reflections allowed Rosalind to calculate key dimensions of the helix with remarkable precision. The work was slow and mathematically demanding, but she believed rigorous analysis was essential before drawing conclusions.

From the image and related calculations, Rosalind determined that:

- DNA is approximately 20 Å in diameter
- The bases are stacked 3.4 Å apart
- One complete helical turn measures 34 Å
- Each turn contains roughly 10 base pairs
- The phosphate backbone lies on the outside of the molecule

The repeating diffraction features also suggested that DNA comprised more than one chain. Her sophisticated cylindrical Patterson analyses of the more challenging A-form provided additional evidence that DNA consisted of two antiparallel strands.

Her calculations revealed that the hydrophilic sugar-phosphate backbone formed the molecule's exterior, with the hydrophobic bases protected inside, a configuration that made chemical and biological sense in the aqueous environment of living cells. This contradicted the first DNA model proposed by Watson and Crick, which had mistakenly placed the phosphates inside the molecule.

Rosalind revealed that the bases in the A-form were slightly tilted and that DNA underwent significant structural changes as it shifted between the hydrated B-form and the drier A-form.

Although Rosalind had not yet determined the precise pairing of the bases, her laboratory notebooks show she was steadily moving toward many of the final structural conclusions. She was already aware of the important chemical work of Erwin Chargaff and Jerry Donohue and was combining these insights with her own experimental evidence. Her meticulous research provided the essential experimental foundation upon which the double-helix model would soon be built, and it represented some of the most advanced structural analyses ever applied to a biological molecule.

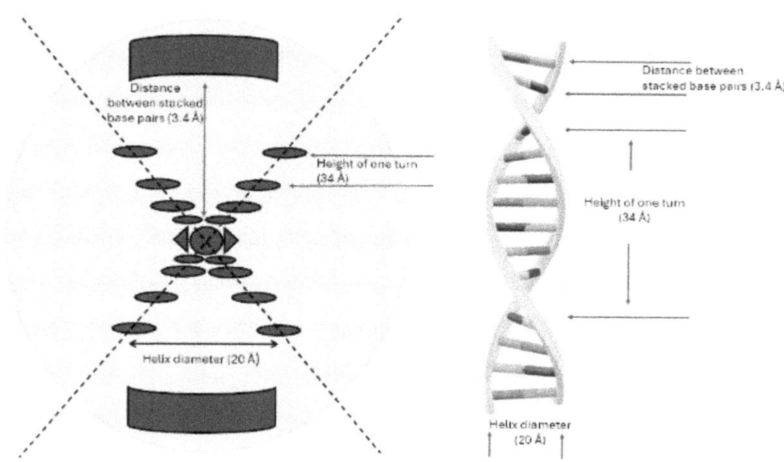


Figure 3. Simplified illustration showing the relationship between an X-ray diffraction pattern and the DNA double helix. Rosalind Franklin's precise measurements of the helix diameter (20 Å), the spacing between stacked base pairs (3.4 Å), and the repeating diffraction pattern provided crucial evidence for determining DNA's structure. The characteristic X-shaped pattern revealed that DNA has a helical structure. Original illustration by the author.

## The "Death of the Helix" and the A-Form of DNA

By 1952, disagreements over DNA's structure had intensified at King's. Wilkins strongly believed DNA was helical, even though he lacked definitive proof. Rosalind, however, trusted evidence over assumption. She recognized that the wetter B-form produced the unmistakable X-shaped pattern of a helix, while the drier A-form generated a dense, complex array of reflections requiring advanced cylindrical Patterson analysis.

As she worked through the difficult A-form data in the spring and summer of 1952, Rosalind briefly questioned whether this form was helical. In a rare moment of humor, she and Gosling created a mock funeral notice announcing the "death" of the helix, aimed partly at Wilkins, whose certainty often irritated her. Despite later misinterpretations, her notebooks and Gosling's recollections show she still accepted the helicity of the B-form; her doubts applied only to the more challenging A-form.

Rosalind focused on the A-form because, as a crystallographer, she valued its sharper, more ordered patterns. The B-form was blurrier but clearly helical, with phosphates on the outside. The A-form remained stubbornly complex, and she refused to draw conclusions without rigorous evidence.

By early 1953, she had begun to conclude that both A- and B-forms were likely helical and composed of two strands. Crick later admitted that her detailed A-form data would have confused him if he had seen it earlier, remarking that he was "glad" he had not examined it before they solved the structure.

# The Race to the Double Helix

## Maurice Wilkins and James Watson

Rosalind focused on the more difficult A-form of DNA rather than the clearer B-form captured in Photo 51. As a meticulous crystallographer, she viewed the photograph as just one piece of evidence among many, prioritizing exact experimental analysis over constructing models.

Photo 51 was kept in the shared X-ray laboratory area at King's. Gosling, who had originally been Wilkins's student, showed it to Wilkins informally. Wilkins believed he still had legitimate access to Gosling's work, unaware that Rosalind considered the image part of her own research program.

On January 30, 1953, Watson visited King's. He and Crick were trying to figure out DNA's structure by building physical models at the Cavendish Laboratory. During the visit, Wilkins showed Watson Photo 51 without Rosalind's knowledge or permission. Watson immediately grasped its significance, later recalling in *The Double Helix*: "The instant I saw the picture my mouth fell open and my pulse began to race."

The X-shaped diffraction pattern strongly suggested a helix to scientists who were already searching for a structural model. Wilkins also shared key measurements from Rosalind's work, including repeating distances within the molecule, which provided further clues for Watson and Crick.

## The MRC Report and the Cambridge Breakthrough

In December 1952, Rosalind submitted a progress report to the Medical Research Council (MRC) containing precise measurements of DNA's structure, including the spacing of phosphate groups and the dimensions of the repeating unit. Her findings strongly supported the idea that the phosphate backbone lay on the outside of DNA, correcting earlier incorrect models proposed by Watson, Crick, and Pauling.

At King's, Rosalind continued advancing through painstaking experimental work. Unlike Watson and Crick, who relied heavily on model building, she approached the problem through rigorous X-ray diffraction analysis and careful mathematical calculations, preferring evidence over speculation.

Rosalind focused on the more ordered A-form of DNA, which produced sharper, more detailed diffraction images than the wetter B-form. However, the complexity of the A-form patterns made them difficult to interpret and temporarily obscured the underlying helical structure.

Her data was far more precise than earlier studies by William Astbury and Florence Bell. Rosalind showed that the phosphate backbone lay on the outside and identified key dimensions of the helix, including the 34-angstrom repeat and the 20-angstrom diameter. She also determined the DNA space group C2 symmetry, showing an even number of strands running in opposite directions. Because Crick had studied this symmetry in his hemoglobin work, he quickly recognized that DNA's strands must run antiparallel.

Although Rosalind knew that both King's and Cambridge were studying DNA, she did not know that her unpublished findings would soon be shared with Watson and Crick.

By early 1953, the race to determine DNA's structure had intensified. Watson and Crick were eager to solve the puzzle before Linus Pauling corrected his own flawed triple-helix model, which incorrectly placed the phosphate groups at the center. Watson and Crick had previously made a similar mistake.

Max Perutz, head of the MRC Unit at the Cavendish Laboratory, routinely received the annual reports circulated among MRC laboratories, including the King's Biophysics Unit. In February 1953, believing the report to be a public administrative document, he showed Rosalind's section to Watson and Crick. The report was not officially confidential, and Perutz did not think he was doing anything improper. However, Rosalind had assumed the report would remain internal to King's.

The information proved crucial. Crick immediately recognized that the symmetry described in Rosalind's report pointed to two antiparallel strands rather than three. Combined with Chargaff's base-pairing rules and Jerry Donohue's advice on the correct chemical forms of the bases, Watson and Crick suddenly had many of the missing pieces needed to solve the structure.

## February 28, 1953

Watson and Crick's breakthrough depended heavily on experimental evidence produced at King's, particularly Rosalind's famous X-ray diffraction image known as Photo 51. Using the measurements and structural clues from Photo 51, along with Rosalind's precise analysis of DNA's dimensions, Watson and Crick constructed their physical model with metal pieces and used cardboard cutouts for the bases. Through trial and error, they could quickly shift the cardboard cutouts of the four bases into configurations that matched the available evidence. Using Rosalind's measurements and their own chemical reasoning, they finally assembled the correct double-helix structure.

On February 28, 1953, Watson and Crick determined that the DNA structure was a double helix. The structure consisted of two long strands running in opposite directions and twisting around one another like a spiral staircase. Paired organic bases, adenine with thymine and cytosine with guanine, were between the strands.

Later that day, at lunchtime in the Eagle pub in Cambridge, Crick reportedly announced to the room, "We have discovered the secret of life." The statement marked one of the most significant scientific breakthroughs of the twentieth century.

## DNA Double Helix

By early March 1953, Watson and Crick had assembled a model that fit the available evidence: two antiparallel strands twisted into a double helix, with adenine pairing with thymine and cytosine with guanine to form the molecule's interior. The structure explained both the uniform width of the helix and how DNA could replicate, with each strand serving as a template for its complement.

Although Watson and Crick built the final model, it depended heavily on dimensions revealed by Rosalind's X-ray work, especially Photo 51 and her unpublished measurements of the helix repeat and molecular symmetry. They later insisted they were "not aware of the details," claiming the structure rested mainly on published data and stereochemical reasoning. Rosalind, however, did not know how extensively her results had guided their work until after the model was complete.

Meanwhile, Rosalind was still analyzing DNA experimentally. Her notebooks show she was close to many of the same conclusions: the helical nature of the molecule, the placement of the phosphate groups on the outside, and the significance of the repeat spacing. She simply had not yet taken the final conceptual step when Watson and Crick announced their structure.

In March 1953, Watson and Crick invited both Rosalind and Wilkins to Cambridge to see the finished model. Rosalind, who had independently reached many of the same insights only days earlier, agreed that the structure should go forward to publication.

---

**SCIENCE SPOTLIGHT**

### The DNA Double Helix

The double helix is the three-dimensional structure of DNA, the molecule that carries genetic information in all living organisms. It comprises two long strands of nucleotides twisted around each other in a spiral, forming a stable yet flexible architecture for storing biological instructions.

A sugar-phosphate backbone forms each strand, with projecting bases: adenine (A), thymine (T), cytosine (C), and guanine (G). The bases pair in a strict pattern (A with T, C with G), creating "rungs" that hold the two strands together. Known as antiparallel orientation, the strands run in opposite directions.

The helical shape reflects the geometry of the sugar-phosphate backbone and the stacking of the bases, which naturally align in neat coin-like layers. This arrangement stabilizes the molecule and protects the genetic code at its center.

The double helix provides a mechanism for accurate copying. When cells divide, the two strands separate, and each becomes a template for building a new partner strand. This complementary base-pairing mechanism ensures that genetic information is passed on with remarkable fidelity.

The discovery of the double helix in 1953 transformed biology. It revealed how heredity works at the molecular level, explained mutation and variation, and laid the foundation for modern genetics, biotechnology, and the Human Genome Project.

## Rosalind's Resignation

By the beginning of 1953, Rosalind may have been closer than anyone else to solving the structure of DNA through experimental evidence alone. Rosalind approached the problem with mathematical rigor, unlike Watson and Crick, who relied heavily on physical model building and insisted that precise X-ray diffraction data grounded every conclusion.

Rosalind was making important progress in her understanding of DNA. However, the complicated A-form had resisted interpretation for many months. Although Rosalind still focused primarily on the A-form, she also began a new laboratory notebook devoted to the B-form before leaving

King's, because the B-form presented a simpler and more clearly helical structure.

Her unpublished drafts and laboratory notes from late February 1953 show that she had accepted DNA as helical, likely with two strands, and that she was positioning the bases in ways that would allow the molecule to replicate. She had also recognized that the order of the bases along a strand did not affect the overall structure of the molecule. As she wrote, "an infinite variety of nucleotide sequences would be possible to explain the biological specificity of DNA." This was a major insight into how DNA could carry genetic information. Years later, her colleague Aaron Klug remarked Rosalind had been "two steps away" from solving the structure herself.

However, in March 1953, Rosalind formally resigned from King's, as she had become deeply unhappy in the environment there. Her difficult working relationship with Wilkins, combined with the broader atmosphere within the department, convinced her to accept a new position at Birkbeck College in London, where she would move into virology research.

As part of her transfer to Birkbeck, Rosalind was required to leave all her DNA research behind at King's. King's also did not permit her to continue the project, and Wilkins assumed responsibility for the work. As part of the transition, Gosling also returned to working under Wilkins's supervision.

For Rosalind, however, leaving King's was more important than remaining in a difficult and increasingly unhappy environment. After two difficult years of her three-year fellowship at King's, she was relieved to move on to the next chapter of her scientific career.

## 1953 Nature Paper

In mid-March 1953, Wilkins and Rosalind visited Cambridge to see Watson and Crick's DNA model and immediately agreed it was correct. They decided the model would appear as Watson and Crick's work alone, with the King's researchers publishing their supporting X-ray data separately.

On 18 March, Wilkins acknowledged receiving the Watson and Crick manuscript. A day earlier, Rosalind, preparing to leave King's for Birkbeck, had completed a draft outlining her conclusions about the B-form backbone. She made only minor edits before submitting her April 1953 *Nature* paper, which appeared third in the trio and which others often interpreted as merely confirming the Cambridge model.

Watson and Crick's announcement offered no experimental evidence beyond Photo 51, the B-form diffraction pattern taken by Rosalind and Gosling. Watson later wrote in *The Double Helix* that the Cambridge team had been willing to "cut corners," though many contemporaries denied any sense of a race and shared Rosalind's caution about premature conclusions.

The April 1953 *Nature* issue published three consecutive papers: Watson and Crick's double-helix model; Wilkins, Stokes, and Wilson's X-ray data; and Rosalind and Gosling's analysis. Watson and Crick noted that base pairing "immediately suggests a possible copying mechanism," later called "the secret of life," and acknowledged being "stimulated" by unpublished King's results only in a footnote.

This limited credit reinforced the impression that King's contributions were secondary. Although Randall and the Cavendish director arranged the paper order, it shaped public perception. Rosalind and Gosling's 'Molecular Configuration in Sodium Thymonucleate' lacked the dramatic framing of the Cambridge announcement and made less of an immediate impact.

Watson and Crick never told Rosalind they had seen her data, nor did they explicitly acknowledge relying on it. Cambridge and parts of King's excluded her from their informal exchanges. While some objected to the Cambridge team's use of King's data, senior Cavendish figures viewed it as normal practice. There is no evidence Rosalind objected, though Watson and Crick could have sought permission and made their dependence clear.

This borrowing, and the limited credit she received, became the root of one of science's most enduring disputes. A draft *Time* magazine article written "in consultation with Franklin," though never published, described the work as joint. A colleague's letter suggests Rosalind knew Crick was seeing some of her results, but the boundaries around data sharing were never clearly defined.

Rosalind often said that "the facts should speak for themselves." Her insistence on solid evidence strengthened her record but made her appear reluctant to speculate. She left King's months before the *Nature* papers appeared and did not attend the April 25, 1953, celebration. By then, she had settled at Birkbeck and had moved on from DNA to new lines of research.

At the Royal Society Conversazione in June 1953, the authors of all three *Nature* papers signed one of the first public displays of the double helix.

## Rosalind's Drafts

By January 1953, Rosalind had resolved many contradictions in her data. She concluded that both A- and B-form DNA were helical, likely composed of two intertwined strands with the phosphate backbone on the outside, and began drafting papers outlining these findings.

Two manuscripts on the A-form were submitted to *Acta Crystallographica* on March 6, 1953, one day before Watson and Crick completed their double-helix model. She had clearly written and mailed them before knowing the Cambridge structure. On July 8, 1953, she revised one paper "in proof" to reflect the newly published results from both King's and Cambridge.

Years later, Aaron Klug discovered a third unpublished draft on the B-form, dated March 17, 1953, among her Birkbeck papers. He noted how closely her conclusions aligned with those in the April 1953 *Nature* papers, showing she was extremely close to solving the structure independently.

Three weeks after the *Nature* papers appeared, Sir Lawrence Bragg lectured on the discovery at Guy's Hospital Medical School. Journalist Joan Bruce attended and began preparing a *Time* magazine article portraying the breakthrough as the work of "two teams": Rosalind and Wilkins gathering experimental evidence, and Watson and Crick focusing on theoretical interpretation. She emphasized Rosalind as an equal contributor and noted that Watson and Crick pursued the double helix after examining King's X-ray data, while Rosalind continued testing the Cambridge model against her own results. The article was never published, possibly because Rosalind felt the science needed further clarification.

Debate later continued over how much Watson and Crick relied on Rosalind's unpublished work. In a 1954 paper, they acknowledged that without her data, "the formulation of our structure would have been most unlikely, if not impossible." They also recognized that Rosalind and Wilkins had independently suggested that B-form DNA was helical, composed of two intertwined chains with the phosphate backbone on the outside.

## Virus Research at Birkbeck College

When Rosalind arrived at Birkbeck College in 1953, she entered a scientific environment very different from the tense and often uncomfortable atmosphere she had experienced at King's. At Birkbeck, she joined the laboratory of the distinguished crystallographer John Desmond ('J. D.') Bernal, a pioneer of molecular biology and X-ray crystallography. Bernal immediately recognized Rosalind's extraordinary technical ability and encouraged her to apply her expertise to one of the newest and most challenging problems in science: the structure of viruses.

For Rosalind, the move marked both a professional and personal turning point. Free from many of the tensions that had plagued her at King's, she quickly established herself as a leading figure in structural virology. Although her role in the discovery of DNA would later dominate public discussion of her career, many scientists believed her virus research was equally important, and in some ways even more original.

At the time, remarkably little was known about the physical structure of viruses. Scientists understood viruses caused disease, but they did not yet fully understand how viruses assembled themselves, reproduced, or infected cells. Rosalind realized that X-ray crystallography could provide answers to these questions by revealing the precise arrangement of molecules within virus particles.

Rather than beginning with human viruses, Rosalind focused initially on plant viruses, which were easier to isolate, purify, and study in large quantities. Her primary subject became the tobacco mosaic virus (TMV).

## Tobacco Mosaic Virus

TMV was one of the best-known viruses in the world during the 1950s. It infected tobacco plants aggressively and spread rapidly through crops, producing a mottled or "mosaic" discoloration on leaves that gave the virus its name. TMV had become scientifically important because it was one of the first viruses ever discovered and one of the few that researchers could reliably purify and examine in the laboratory.

The virus itself was extraordinarily small, far beyond the limits of ordinary microscopes. Scientists already knew that TMV consisted mainly of two parts: genetic material in the form of ribonucleic acid (RNA) and an outer protein coat. However, they did not know the arrangement of these components inside the virus particle. One major question concerned the location of the RNA. Some scientists believed it sat loosely within a hollow center, while others suspected it was more closely integrated with the surrounding proteins. Rosalind set out to solve this problem using X-ray diffraction.

## Applying X-Ray Crystallography to Viruses

The techniques Rosalind had mastered while studying coal and DNA could also apply to viruses, but the work was even more difficult. Viruses are highly complex biological particles made of repeating protein units, and researchers had to carefully purify and align these units before obtaining meaningful diffraction patterns.

Rosalind approached the problem with the same meticulous precision that had characterized all of her earlier work. She improved methods for preparing virus samples, refining techniques for drying, orienting, and exposing virus particles to X-rays. Her goal was to produce clearer and more detailed diffraction images than any previously obtained.

Working with members of her research group, including Klug and Kenneth Holmes, Rosalind gradually generated increasingly sophisticated diffraction photographs of TMV. From these patterns, she began calculating the virus's dimensions and internal organization.

Her work showed that TMV possessed a helical structure. The protein subunits formed a spiral around the virus, rather than stacking in simple layers. This was a breakthrough in virology.

Rosalind also showed that the virus's RNA was not sitting loosely inside an empty tube, as many researchers had believed. Instead, the RNA embedded itself within the inner surface of the protein coat and followed the helical structure of the virus itself. This discovery fundamentally changed scientists' understanding of virus organization.

## Understanding the Structure of TMV

Rosalind's measurements revealed that thousands of repeating protein molecules arranged in a highly ordered helical pattern around a single strand of RNA constructed TMV. She showed that:

- The virus had a rod-like shape,
- The protein subunits repeated regularly,
- The RNA followed the inner groove of the helix,
- And the overall structure was both stable and highly organized.

Her work helped explain one of the central mysteries of virology: how viruses could self-assemble.

Scientists began to understand that virus particles did not require an outside force to construct them. Instead, the protein molecules and RNA could naturally organize themselves into functional virus particles through ordinary chemical interactions. This became one of the foundational ideas of molecular virology.

Rosalind also showed that viruses could undergo structural changes depending on environmental conditions such as humidity and temperature. These observations helped researchers better understand viral stability, infection, and survival outside host organisms.

## Collaboration with Aaron Klug

One of Rosalind's most important scientific partnerships at Birkbeck was with Aaron Klug, a young researcher who joined her laboratory shortly after her arrival. The two worked closely on virus structure and diffraction analysis, combining X-ray crystallography with emerging electron microscopy techniques.

Klug later spoke frequently about Rosalind's scientific discipline, technical brilliance, and insistence on rigorous evidence. He admired her ability to extract meaningful conclusions from difficult experimental data while refusing to make unsupported claims. Together, Rosalind and Klug produced increasingly accurate structural models of viruses. Their work helped establish many of the methods later used in structural molecular biology.

Years after Rosalind's death, Klug received the 1982 Nobel Prize in Chemistry for work that directly built upon the techniques and concepts developed during their collaboration at Birkbeck. Throughout his career, Klug consistently acknowledged Rosalind as one of the most important scientific influences in his life.

## Expanding Beyond TMV

Following the success of her TMV research, Rosalind expanded her studies to several other plant viruses, including potato virus X, turnip yellow mosaic virus, cucumber viruses, and tomato bushy stunt virus. Under her leadership, the Birkbeck Laboratory quickly became one of the world's leading centers for virus structure research.

Rosalind also began studying poliovirus, the virus responsible for polio, one of the most feared diseases of the era. At the time, vaccines were still under development, and scientists believed that understanding the physical structure of the virus would be essential for future medical advances. Rosalind's team used X-ray diffraction to investigate the size, symmetry, and protein shell of the poliovirus particle, aiming to determine whether it shared the same icosahedral architecture seen in many plant viruses. She worked closely with Klug and other colleagues to refine methods for preparing and imaging the virus, helping establish experimental approaches that later became standard in structural virology.

Although Rosalind died before completing this work, her early findings, particularly her evidence for poliovirus's symmetrical, protein-based outer shell, guided later researchers who successfully solved its structure. Her contributions laid the foundation for understanding how human viruses assemble, infect cells, and how vaccines can target them, extending her influence far beyond the plant viruses that first made her reputation.

---

**SCIENCE SPOTLIGHT**

## Polio and Rosalind Franklin's Findings

Polio is a highly infectious viral disease that once caused widespread paralysis, especially in children. Before scientists developed effective vaccines in the 1950s, outbreaks created intense public fear, and they raced to understand the virus at a structural level.

At Birkbeck College, Rosalind applied X-ray diffraction to poliovirus, developing methods to purify the virus and capture clear, interpretable diffraction patterns. Her work provided early evidence that poliovirus particles were small, highly symmetrical, and built from repeating protein subunits arranged in an icosahedral shell. This was one of the first demonstrations that a human virus could share the same geometric architecture seen in many plant viruses.

These findings helped establish the structural principles that later researchers used to solve the full poliovirus structure and informed how scientists approached viral assembly, infection, and vaccine design. Although Rosalind did not live to complete the project, her experimental strategies and early insights laid the foundations that modern structural virology still builds upon.

---

## A New Scientific Field

Rosalind's virus research helped transform virology from a largely descriptive science into a structural and molecular one. Before Rosalind's work, scientists often studied viruses mainly through the diseases they caused in plants, animals, or humans. Instead, Rosalind treated viruses as highly organized molecular structures that she could measure mathematically and understand at the atomic level.

Her work showed that:

- Viruses possessed precise geometric organization,
- Biological molecules could self-assemble into highly ordered structures,
- And X-ray crystallography could reveal the architecture of living systems at the molecular scale.

These ideas became central principles of modern molecular biology and structural virology.

Today, many historians of science believe that Rosalind's contributions to virus research are just as significant as her work on DNA. Although Photo 51 became the most famous image associated with her name, her studies at Birkbeck helped lay the foundations for modern structural biology, virology, and even aspects of modern vaccine research decades later. Rosalind helped establish structural virology, a field that remains central to understanding viruses, vaccine development, and infectious disease research today.

---

**SCIENCE SPOTLIGHT**

### Structural Biology and Vaccines

Structural biology is the field that reveals the three-dimensional shapes of biological molecules. In virology, it shows how viruses are built, how their protein shells assemble, how they attach to cells, and how they change shape during infection. These structural insights are essential for designing vaccines.

By determining the precise arrangement of proteins on a virus's surface, scientists can identify the regions most exposed to the immune system. Knowing the exact shape of these targets allows researchers to design vaccines that present the immune system with the right molecular features, prompting a strong and specific response. Structural biology also explains why some viruses mutate rapidly, why certain variants evade immunity, and how updated vaccines can counter them.

---

## Recognition at Last

Unlike the difficult years at King's, Rosalind's time at Birkbeck brought her growing international recognition. She lectured widely, collaborated with scientists around the world, and became respected as one of the leading experts in virus structure research.

Her colleagues remembered her laboratory as energetic, intellectually demanding, and collaborative. Rosalind pushed her team hard, but she also inspired enormous loyalty and admiration. By the mid-1950s, she was no longer simply a scientist associated with DNA. She had become one of the pioneers of structural virology.

Today, many historians argue that Rosalind's virus research was every bit as important scientifically as her work on DNA, and in some ways showed even more clearly the full range of her brilliance as an experimental scientist.

## Illness and Final Years

Although Rosalind finally found professional happiness at Birkbeck, tragedy cut her life short just as her virus research was gaining international recognition.

During her years at Birkbeck, she traveled twice to the United States, in 1954 and 1956, and deeply appreciated the respect she received there. American scientists regarded her as a leading expert in X-ray crystallography and virus structure, a welcome contrast to the strained atmosphere she had experienced at King's.

During her 1956 visit, Rosalind began suffering from severe abdominal pain and swelling. She ignored the symptoms at first, but after returning to London, doctors diagnosed ovarian cancer. She underwent surgery in September 1956, followed by radiation and experimental chemotherapy.

Even after her diagnosis, Rosalind refused to abandon her research. Encouraged by colleagues, she had recently begun studying poliovirus, then one of the most feared diseases of the era. She secured a safe laboratory for handling live virus samples and immediately applied her crystallographic methods to the problem.

Her determination astonished those around her. Despite increasing weakness, she continued working throughout 1957, often returning to the lab between hospital stays. Friends recalled her arriving exhausted but instantly focusing on diffraction photographs, calculations, and discussions with younger researchers.

Her group made remarkable progress. Using X-ray diffraction and electron microscopy, they showed that poliovirus possessed icosahedral symmetry, a highly organized structure with twenty triangular faces and sixty repeating subunits. This became an important step toward understanding viral assembly and infection.

Rosalind continued supervising students, writing papers, securing funding, and expanding her virus program. In 1956 and 1957 alone, her laboratory produced more than a dozen papers, an extraordinary achievement given her illness.

During periods of recovery, she often stayed with friends or family. Crick and his wife, Odile, welcomed her into their home during treatment, reflecting the warm friendship that had developed after the DNA discovery.

Biographers have suggested that her cancer may have been linked to long-term X-ray exposure, though this has never been proven. Safety standards in the 1940s and 1950s were far less rigorous, and she frequently worked with powerful equipment without modern protection. There was also a family history of cancer.

Throughout her illness, Rosalind remained intensely committed to science. Colleagues remembered her discipline and courage, noting that she refused self-pity and continued planning experiments, reviewing data, and discussing future projects even as her condition worsened.

## Death

Rosalind died in London on April 16, 1958, at the age of just thirty-seven.

Tragically, she died only one day before the Brussels World's Fair opened, where a massive model of the tobacco mosaic virus she had designed received international acclaim.

After her death, colleagues completed and published several of her unfinished virus studies. Collaborators, including Klug and John Finch, who dedicated the resulting publications to her memory, continued her work on poliovirus. Decades later, Klug would receive the 1982 Nobel Prize in Chemistry for work on protein–nucleic acid structures that had begun during his collaboration with Rosalind.

Bernal, who had supported Rosalind throughout her years at Birkbeck, wrote movingly after her death: "As a scientist, Miss Franklin was distinguished by extreme clarity and perfection in everything she undertook. Her photographs are among the most beautiful X-ray photographs of any substance ever taken."

Rosalind was buried at Willesden Jewish Cemetery in northwest London. Her tombstone bears the simple inscription: "Scientist. Her research and discoveries on viruses remain of lasting benefit to mankind.

The wording is fitting. Although Rosalind is most often remembered for her contribution to discovering the structure of DNA, she also helped lay the foundations of modern structural virology. Her work transformed scientific understanding of viruses and showed the power of X-ray crystallography in revealing the hidden architecture of life itself.

## Nobel Prize

When the Nobel Committee awarded the 1962 Nobel Prize in Physiology or Medicine to Watson, Crick, and Wilkins for discovering the structure of DNA, Rosalind's name was absent. She had been dead for four years, and the Nobel Foundation does not award posthumous prizes. Even so, her omission remains one of the most debated issues in modern science.

At the time, none of the laureates substantially acknowledged how essential Rosalind's experimental work had been. Wilkins briefly mentioned her X-ray studies, but Watson and Crick made little direct reference to her contributions in their Nobel lectures. Privately, however, Crick later admitted that "the data which really helped us to obtain the structure was mainly obtained by Rosalind Franklin."

Historians cannot know whether Rosalind would have received the prize had she lived, though many believe she would have been a strong candidate. Because no more than three people can share Nobel Prizes, one of the eventual recipients would have had to exclude someone. Some argue she deserved recognition alongside Watson and Crick; others believe she might instead have earned a Nobel Prize in Chemistry for her later virus research.

What is beyond dispute is that her work was fundamental. Her X-ray diffraction images, especially Photo 51, provided crucial evidence that DNA was helical. Her measurements established key dimensions, confirmed that the phosphate backbone lay on the outside, and revealed the symmetry supporting the double-helix model. Her notebooks show she had independently reached several conclusions close to Watson and Crick's final structure.

Many historians now believe Rosalind came extremely close to solving the structure herself. Her cautious, evidence-driven approach differed sharply from Watson and Crick's rapid model-building style. She preferred to verify every conclusion rigorously before publishing.

The debate over her role also reflects the scientific culture of the 1950s, when information was often circulated informally, and boundaries around unpublished data were poorly defined. A 2023 *Nature* article argued that the popular "stolen data" narrative oversimplifies a more complex and collaborative environment, while still emphasizing that Rosalind's contribution was indispensable and insufficiently recognized at the time.

Recognition of her work grew after her death. Her Birkbeck colleague Klug became one of her strongest advocates, helping publish her unfinished research and highlighting her role in DNA discovery. In 1982, when Klug received the Nobel Prize in Chemistry for work that began during their collaboration, he paid tribute to her: "Had her life not been cut tragically short, she might well have stood in this place on an earlier occasion."

Today, molecular biology fully acknowledges Rosalind's central role. Her work not only helped reveal the structure of DNA but also laid key foundations for structural virology and modern biophysics. A contribution once treated as supporting is now recognized as essential to one of the twentieth century's greatest scientific breakthroughs.

## Rosalind Remembered by Friends

Friends and colleagues remembered Rosalind as intelligent, lively, and deeply engaged with the world. Friends recalled her enthusiasm for adventure, even when it meant uncomfortable travel or demanding mountain hikes, preferring modest accommodation and valuing experience over luxury. She enjoyed music and theater and was deeply loyal. Away from the lab, she could be playful and humorous, though that side rarely appeared professionally. Those who knew her well described her as warm, witty, and great fun.

However, Rosalind could also be intense, stubborn, and impatient, especially when confronted with poor thinking or carelessness. Several colleagues found her difficult, particularly in the tense atmosphere at King's. Anne Sayre wrote that "absurdities exasperated her," and others described her "fierce and stubborn indignation." One colleague said, "You had to argue strongly with Rosalind," while another called her "dedicated, driven, passionate."

Much of this reputation reflected the environment at King's: rigid hierarchies, poor communication, and gender bias. Rosalind demanded high standards and often clashed with colleagues unaccustomed to outspoken women in senior scientific roles. Later accounts suggest that those who knew her less well exaggerated these traits. Letters preserved by family and friends reveal frustration during difficult periods, but also curiosity, affection, and enthusiasm for science. Even during her unhappy years at King's, her focus remained on the work.

One well-known story involved a dispute over a Tesla coil used for X-ray experiments. After borrowing it from a younger researcher named Brown and failing to return it promptly, Rosalind simply removed it again when he reclaimed it. Brown later recalled the incident with amusement, and the two became close friends.

## Legacy

In the decades after her death, Rosalind's contributions gradually received the recognition they deserved. Many universities, laboratories, awards, and scientific programs now bear her name, preserving her legacy and inspiring future generations.

In 2004, the Chicago Medical School became Rosalind Franklin University of Medicine and Science. Many academic buildings and research centers also honor her, including the Franklin-Wilkins Building at King's and the Rosalind Franklin Institute in Britain.

The Royal Society established the Rosalind Franklin Award to recognize outstanding contributions by women in science, technology, engineering, and mathematics, reflecting both her achievements and the barriers she faced in mid-twentieth-century science.

In 2013, a commemorative plaque was installed at The Eagle pub in Cambridge, where Watson and Crick announced they had discovered "the secret of life."

Her influence extends far beyond DNA. In 2019, the European Space Agency named its ExoMars rover after her, a fitting tribute for a scientist whose work transformed our understanding of life at the molecular level.

Rosalind's later research on viruses also proved enormously important. Her pioneering studies of helical and spherical viruses helped establish structural virology and influenced later work on vaccines and molecular biology. Scientists note that her insights remain relevant to modern virology, including studies of coronaviruses and other emerging pathogens.

## Why She Matters Today

Rosalind's research helped reveal the structure of DNA, one of the most important scientific discoveries of the twentieth century. Today, our understanding of DNA underpins modern genetics and medicine, enabling scientists to diagnose inherited disorders, trace ancestry, study evolution, and develop targeted therapies. The structural insights that Rosalind and her colleagues first made possible enabled landmark achievements such as the Human Genome Project.

Her contributions to virus research are equally significant. By applying X-ray crystallography to viruses, she helped establish the foundations of structural virology, providing techniques that continue to guide scientists in understanding viral architecture, developing vaccines, and combating infectious diseases. Although she did not live to see these advances, her work helped shape many of the tools used in modern medicine.

Rosalind's meticulous approach to science, careful sample preparation, controlled experiments, and rigorous interpretation of evidence set standards that remain central to structural biology today. Her insistence on accuracy over speculation continues to influence how researchers investigate complex biological systems.

While Photo 51 became the image most closely associated with her name, her legacy extends far beyond a single photograph. Her pioneering studies of carbon materials advanced industrial chemistry; her experimental work provided crucial evidence that made the DNA double helix model possible, and her research on RNA viruses laid the groundwork for modern structural virology.

Rosalind's career also highlights the importance of fairness, collaboration, and proper attribution in scientific research. Her story remains a powerful reminder that scientific progress depends not only on brilliant ideas but also on recognizing the contributions of every researcher. In an age of rapid information sharing and artificial intelligence, her commitment to rigorous evidence before drawing conclusions is more relevant than ever.

Today, Rosalind continues to inspire scientists around the world, particularly women entering fields where recognition has not always been equitable. Her legacy is not only the discoveries she made but also the example she set: that careful observation, intellectual honesty, and perseverance can transform our understanding of the natural world.

In the end, Rosalind's work spoke exactly as she believed it should: through evidence, through truth, and through the lasting clarity she brought to the hidden architecture of life. She transformed carbon science, provided experimental evidence that made the DNA double helix possible, pioneered structural virology, and helped establish the techniques that underpin modern molecular biology. Her influence continues to shape science and medicine in ways that touch millions of lives every day.

Today, Rosalind is remembered not simply as the scientist behind Photo 51, but as a pioneer of structural biology whose meticulous research transformed our understanding of DNA, viruses, and the molecular architecture of life itself.

## Timeline

| Year | Event |
|---|---|
| 1920 | Born in London |
| 1931 | Enrolled at St. Paul's Girls' School |
| 1938 | Entered Newnham College, Cambridge |
| 1941 | Completed Natural Sciences Tripos at Cambridge |
| 1942 | Joined BCURA |
| 1945 | Awarded her PhD from Cambridge |
| 1946 | Published her first peer-reviewed paper |
| 1947 | Moved to Paris |
| 1948 | Officially received a Bachelor of Arts degree from Cambridge |
| 1951 | Joined King's College London |
| 1952 | Photo 51 taken |
| 1953 | Watson & Crick announced the DNA double helix; *Nature* journal DNA edition published; Rosalind joins Birkbeck College, London |
| 1954-1956 | Toured the U.S. |
| 1957 | Expanded virus studies |
| 1958 | Died at age 37 |
| 1962 | Nobel Prize awarded to Watson, Crick, and Wilkins |
| 2000s | Rosalind's work recognized |

"Science and everyday life cannot and should not be separated."

—*Rosalind Franklin*

# VERA RUBIN

(1928-2016)

## Astronomer

Her discoveries revealed compelling evidence for the existence of dark matter.

Original portrait illustration by Osmo, based on a historical reference photograph (c. unknown)

"Science progresses best when observations force us to alter our preconceptions."
—*Vera Rubin*

When women often faced discouragement from looking through telescopes, Vera Rubin entered astronomy, let alone challenging established theory. Still, she followed her fascination with the motion of galaxies wherever it led. Her discoveries would reveal one of the universe's deepest mysteries and reshape modern cosmology.

## Childhood

Vera Florence Cooper was born on July 23, 1928, in Philadelphia, Pennsylvania. She was the younger of two daughters born to Philip and Rose Cooper, both of whom came from Jewish immigrant families that had fled Eastern Europe.

Her father, born Pesach Kobchefski in Vilnius, Lithuania, immigrated to the U.S. with his family when he was seven years old to escape the persecution and instability faced by Jewish communities in Eastern Europe. After arriving in America, the family changed their surname to Cooper. Vera's mother, Rose Applebaum Cooper, was born in the U.S. after her parents emigrated from Bessarabia, a region that today includes parts of Moldova and Ukraine. Philip, an electrical engineer, and Rose met while employed at the Bell Telephone Company in Philadelphia. Their first daughter, Ruth, was born several years before Vera.

In 1938, when Vera was ten years old, the family moved to Washington, D.C., after Philip accepted a position with the United States Department of Agriculture. The move would prove significant. It was in Washington, D.C., that Vera's lifelong fascination with the night sky took shape.

From an early age, Vera displayed an intense curiosity about the world around her. Unlike many children whose interests changed frequently, Vera became captivated by a single question: what was happening beyond Earth?

At around eleven years old, she began spending evenings looking out of her bedroom window at the stars. The more she observed, the more she wanted to understand. While other children her age were asleep, Vera often remained awake, watching constellations move across the sky and wondering what lay beyond them.

Her parents encouraged this curiosity. Rather than dismissing astronomy as an unusual interest for a young girl, they supported it wholeheartedly. Her father became an enthusiastic partner in her early explorations.

Together, they built a small homemade telescope using cardboard tubing and a modest two-inch Edmund Scientific lens. It was a simple instrument, but for Vera, it opened an entirely new universe. The telescope was only the first of many she would construct during her life.

Whenever she observed a meteor streaking across the sky, she would carefully note its path and attempt to map its trajectory the following morning. Even as a child, she approached observation systematically, recording what she saw and looking for patterns. The habits that would later define her scientific career were already beginning to emerge.

Vera's father also took her to meetings of the D.C. Amateur Astronomers Club. These gatherings introduced her to a community of people who shared her fascination with the heavens. She listened eagerly as amateur and professional astronomers discussed discoveries and debated ideas about the universe.

Through these meetings, she could hear lectures by prominent astronomers such as Harlow Shapley and David Menzel. For a young girl with dreams of becoming an astronomer, the experience was both inspiring and affirming.

---

**HISTORICAL SPOTLIGHT**

## Harlow Shapley and Donald Menzel

Harlow Shapley (1885-1972) was one of the most influential American astronomers of the early 20th century. His pioneering work on globular clusters revealed the true scale of the Milky Way and shifted the Sun from the center of the galaxy to a more modest position on the outskirts. As director of the Harvard College Observatory, Shapley shaped U.S. astronomy for decades, advocating for large surveys, public science education, and international collaboration. People also knew him for his strong opinions and for being willing to challenge prevailing ideas. The scientific community admired and debated these traits.

Donald Menzel (1901-1976) was a leading astrophysicist and one of the world's foremost experts on the Sun. His research on solar spectra, the chromosphere, and the physics of ionized gases helped shape the foundations of modern solar astronomy. As director of the Harvard College Observatory after Shapley, Menzel modernized the institution and expanded its scientific reach. He was also a gifted science communicator, writing popular books that introduced astronomy to wide audiences. Known for his sharp intellect and equally sharp debates, Menzel played a major role in shaping mid-century American astrophysics.

---

Years later, Vera recalled how completely astronomy had captured her imagination:

"By about age 12, I would prefer to stay up and watch the stars than go to sleep. I started learning. I started going to the library and reading. But it was initially just watching the stars from my bedroom that I really did. There was just nothing as interesting in my life as watching the stars every night."

Her enthusiasm soon outgrew the children's section of the local library. Recognizing her determination, her mother persuaded the librarians to allow Vera to borrow books from the adult science collection. She eagerly read everything she could find about astronomy, mathematics, and physics.

As her interest deepened, she began photographing the night sky. Long-exposure photographs often produce streaks of light as stars appear to move across the image. Frequently, photo-processing laboratories assumed the streaks were mistakes and attempted to correct them. Rose repeatedly defended her daughter's photographs, explaining that the unusual patterns were intentional and reflected astronomical observations rather than photographic errors.

The support Vera received at home was unusual for the time. During the 1930s and 1940s, few encouraged girls to pursue careers in science. The Cooper household was different. Philip and Rose expected both of their daughters to pursue whatever careers interested them.

The sisters eventually followed very different paths. Ruth entered the legal profession and later became a judge, while Vera devoted herself to science and the study of the universe.

Looking back, Vera often emphasized that curiosity, rather than ambition, was what drove her. Even as a child, she found herself fascinated by questions more than by answers: "Even then I was more interested in the question than in the answer. I decided at an early age that we inhabit a very curious world."

That curiosity would become the foundation of a scientific career that would ultimately transform our understanding of the universe.

## High School Life

Vera attended Coolidge Senior High School in Washington, D.C., graduating in 1944. By this point, she had already decided that she wanted to become an astronomer, a career choice that was unusual for women at the time and often actively discouraged.

Although she excelled academically, not all her teachers supported her ambitions. Vera later recalled that some educators held deeply sexist views about the role of women in science.

One physics teacher began the school year by dividing scientific discoveries into two categories: those requiring brilliance and insight, which he illustrated only with male scientists, and those requiring diligence and hard work, for which he used Marie Curie as his example. Vera found such comments infuriating. She later remembered:

"The first day of class, he defined two kinds of discoveries: those that took insight and brilliance and those that required hard work but not brilliance. His examples of brilliance were all men. His example of hard work was Marie Curie. I was often angry in class for his similar comments."

The discouragement became even more direct when she announced her intention to pursue science at university. One teacher advised her to abandon the idea entirely, saying, "As long as you stay away from science, you should do OK."

Fortunately, Vera ignored the advice. Rather than allowing these attitudes to discourage her, they strengthened her determination. She had already spent years observing the night sky, reading astronomy books, and attending astronomy club meetings. She knew exactly what she wanted to study.

When the time came to choose a university, Vera selected Vassar College in Poughkeepsie, New York. The decision was deliberate. Vassar was one of the few institutions in the U.S. where women could study astronomy seriously, and it had a distinguished scientific heritage. The college was the alma mater of Maria Mitchell, the first professional female astronomer in the United States and one of Vera's early inspirations.

When opportunities for women in science remained limited, Vera arrived at Vassar determined to pursue astronomy despite the obstacles that lay ahead. It was the first step in a journey that would eventually change humanity's understanding of the cosmos.

## Vassar College

In 1944, Vera enrolled at Vassar College. She had earned a scholarship that likely covered much of her tuition, making it possible for her to pursue the dream she had nurtured since childhood: becoming an astronomer.

When relatively few women entered scientific fields, Vassar offered opportunities that were unavailable at many other institutions. The college had a long tradition of supporting women in science and was renowned for its astronomy program.

Vassar's reputation rested largely on the legacy of Maria Mitchell, America's first professional female astronomer. After discovering a comet in 1847, Mitchell achieved international fame. Later, Vassar appointed her as a professor when it established the college observatory in 1865. For generations of young women interested in astronomy, Mitchell served as proof that scientific achievement was possible despite the barriers facing women. For Vera, Mitchell was more than a historical figure. She was an inspiration.

One of the great privileges of studying at Vassar was the opportunity to work in the same observatory where Mitchell had once conducted her observations. Vera spent countless hours there learning observational astronomy, using telescopes, recording measurements, and deepening her understanding of the universe she had loved since childhood.

Her enthusiasm for astronomy quickly became apparent. Vera was famous for her curiosity, determination, and willingness to devote long hours to her studies. While many students explored a variety of academic interests, she remained firmly focused on astronomy. Years later, she recalled that she had never seriously considered another career.

The years at Vassar provided formal scientific training that transformed childhood fascination into professional ambition. She studied mathematics, physics, and astronomy, building the foundations she would later use to challenge some of the most important assumptions in modern cosmology.

By the time she graduated in 1948, Vera was the only student in her graduating class to receive a bachelor's degree in astronomy. She also earned membership in Phi Beta Kappa, the oldest academic honor society in the U.S., recognizing her outstanding academic achievement.

Although graduating as the sole astronomy major might have discouraged some students, Vera saw it as confirmation that she had chosen the path she wanted to follow.

---

**HISTORICAL SPOTLIGHT**

## Maria Mitchell (1818–1889)

Maria Mitchell was the first professional female astronomer in America and a key figure in 19th-century science education. Born on Nantucket to a Quaker family that championed learning for girls, she developed exceptional observational skills under her father's guidance. She discovered a comet in 1847 and became internationally famous, even receiving a medal from the King of Denmark.

Mitchell later joined Vassar College, then an all-women's institution, becoming a professor of astronomy from the founding of its observatory in 1865. Her presence made Vassar famous for its association with pioneering women in science. Through rigorous teaching and outspoken advocacy, she insisted that women deserved full access to scientific training, professional respect, and intellectual independence.

## Barriers to Graduate Education

After earning her degree, Vera planned to continue her studies in astronomy. Like many ambitious young astronomers of the era, she hoped to attend Princeton University in New Jersey, whose Department of Astronomy was among the most prestigious in the country.

However, she encountered a barrier that had nothing to do with her academic ability. Princeton simply did not admit women to its graduate astronomy program. The exclusion was not unique to Princeton. During the 1940s, many universities either prohibited women from enrolling in advanced science programs or severely restricted their opportunities. Princeton would not begin admitting women to its graduate astronomy program until 1975, nearly three decades after Vera's application.

The rejection was an early reminder of the obstacles facing women in science. Throughout her career, Vera would repeatedly encounter discrimination and exclusion. These experiences would later motivate her to become one of astronomy's strongest advocates for women scientists. Fortunately, Vera refused to let the setback derail her ambitions.

During her summers at Vassar, she had already gained valuable scientific experience through positions at the U.S. Naval Observatory and the Naval Research Laboratory in Washington, D.C. These opportunities exposed her to professional astronomical research and strengthened her determination to continue in the field. When Princeton closed its doors, she looked elsewhere.

## Robert Rubin

While attending Vassar, Vera met Robert Joshua Rubin, known to family and friends as Bob. Bob was a graduate student in mathematics at Harvard University and was taking part in the Navy's V-12 College Training Program, a wartime initiative designed to speed up the education of military personnel in fields considered important to the war effort. The two met through mutual friends and quickly developed a close relationship.

Bob shared Vera's intellectual curiosity and love of learning. Their relationship grew while Vera was completing her undergraduate studies, and they soon discovered that they complemented one another remarkably well.

After graduating from Vassar in 1948, Vera married Bob. The newlyweds moved to Ithaca, New York, where Bob was beginning doctoral studies in mathematics at Cornell University. His research focused on topology, a branch of mathematics concerned with the properties of geometric objects and spatial relationships.

For Vera, the move created a new decision. The graduate astronomy program at Harvard had accepted her, a significant achievement for a young woman entering the field. Yet she chose not to attend. Instead, she enrolled at Cornell so that she and Bob could remain together while pursuing their respective academic careers. The decision reflected both her commitment to her marriage and her confidence that she could succeed wherever she studied.

Years later, Vera amused audiences by recounting a letter she received from Donald Menzel, director of the Harvard Observatory. At the bottom of an otherwise formal letter, Menzel had reportedly handwritten: "Damn you women. Every time I get a good one ready, she goes off and gets married."

Intended as a joke, the comment nonetheless mirrored a typical assumption of the period, which was that marriage would inevitably bring a woman's scientific career to a close. Vera would spend the next several decades proving just how wrong that assumption could be.

## Cornell University and a Master's Degree

After graduating from Vassar in 1948, Vera moved to Ithaca, New York, to begin her master's degree at Cornell University, where Bob had begun his doctoral studies in mathematics. Although Cornell was not particularly renowned for astronomy at the time, the department consisted of only a handful of faculty members, but it possessed one of the strongest physics programs in the country.

For Vera, Cornell proved to be an exciting and intellectually stimulating environment. She took much of her coursework through the physics department, which gave her the opportunity to learn from some of the most influential scientific minds of the twentieth century. Among the faculty were physicists Richard Feynman, Hans Bethe, and Philip Morrison, whose lectures exposed students to cutting-edge developments in modern physics. Years later, Vera fondly recalled that "Cornell was a very exciting place because of the physics."

The experience broadened her scientific outlook and strengthened the mathematical foundations that would later prove essential to her astronomical research.

---

### HISTORICAL SPOTLIGHT

#### The Physicists of Cornell University

Philip Morrison, Hans Bethe, and Richard Feynman were central figures in mid-20th-century American physics, each shaping the field in distinct but deeply interconnected ways. Morrison, known for his clarity as both a theorist and educator, helped build the first nuclear reactor and later became a leading voice for scientific responsibility. Bethe, already a renowned theoretical physicist when he arrived at Cornell University, won the 1967

Nobel Prize for explaining the nuclear processes that power stars. Awarded the 1965 Nobel Prize, Feynman, Bethe's brilliant protégé, revolutionized quantum mechanics with a new visual language via his path integral formulation and Feynman diagrams. Together, their work linked wartime physics to the creative explosion of postwar theoretical research, leaving a legacy that shaped generations of scientists.

## Studying the Motions of Galaxies

At Cornell, Vera worked with astronomer Martha Carpenter and became captivated by a fundamental question: How do galaxies move through space? Edwin Hubble had shown in the 1920s that distant galaxies recede from one another as the universe expands, a pattern now known as the Hubble flow. Vera wondered whether this expansion told the whole story.

For her master's thesis, she examined the motions of 109 galaxies, including the Milky Way, searching for patterns beyond simple cosmic expansion, what astronomers now call non-Hubble motions. She asked whether galaxies might share additional, organized motions, much as stars orbit within a galaxy or planets orbit a star.

Her analysis suggested that galaxies exhibited motions that Hubble's law could not fully account for. She proposed that groups of galaxies might be rotating around a common center or axis. Such conclusions challenged prevailing assumptions and were met with considerable skepticism. While later observations proved this specific interpretation wrong, her work offered a crucial insight: galaxies were not randomly distributed through space but instead appeared to be organized into larger structures.

What Vera detected in the late 1940s is now understood as part of the supergalactic plane, a vast, flattened arrangement of nearby galaxy groups that includes the Local Supercluster. However, such ideas were controversial at the time. Her thesis challenged accepted thinking and drew criticism because of the limited data available, yet it also attracted attention by hinting at a far more complex architecture of the universe.

Her thesis would later influence French-American astronomer Gérard de Vaucouleurs, whose mapping of nearby galaxies helped establish the existence of the Local Supercluster and the large-scale structures that define our cosmic neighborhood. In retrospect, Vera's master's research revealed a hallmark of her scientific style: a willingness to follow the evidence, even when it pointed in unexpected directions.

---

**SCIENCE SPOTLIGHT**

## Hubble Expansion and the Hubble Flow

Hubble expansion refers to the discovery that the universe itself is stretching over time. As space expands, galaxies grow farther apart, not because they are flying through space, but because the space between them is increasing. The Hubble flow describes this large-scale motion, which is the smooth outward movement of galaxies as the universe expands and carries them along.

Local gravity dominates regions like galaxy groups and clusters, disrupting the Hubble flow. But on vast cosmic scales, the expansion becomes clear: the greater the distance between two galaxies, the faster they appear to recede from one another.

Hubble's Law, which states that recession velocity increases in proportion to distance, describes this relationship mathematically as $v = H_0 \times d$. Together, Hubble expansion, the Hubble flow, and Hubble's Law form the foundation of modern cosmology, revealing that the universe is dynamic, evolving, and far larger than early astronomers imagined.

---

**SCIENCE SPOTLIGHT**

## The Supergalactic Plane and Gérard de Vaucouleurs

In the 1950s, astronomer Gérard de Vaucouleurs uncovered a remarkable pattern in the distribution of nearby galaxies. He found that the Milky Way, Andromeda, and thousands of neighboring systems lay within a vast, flattened structure he named the Local Supercluster, centered on the Virgo Cluster, rather than being scattered randomly. He discovered this by mapping their positions and brightness.

Within this larger system, de Vaucouleurs identified an even more striking feature: a broad, gently curved sheet of galaxies now known as the supergalactic plane. Many galaxies within roughly 100 million light-years lie along this plane, reflecting the influence of gravity and dark matter on enormous scales. Rather than forming in isolation, galaxies emerge within a cosmic network of filaments and walls, and the supergalactic plane is one of the most prominent of these structures.

To study this architecture, de Vaucouleurs introduced the supergalactic coordinate system, aligning astronomical maps with the true large-scale layout of nearby galaxies. His work provided some of the earliest clear evidence that the universe possesses structure far beyond individual galaxies, revealing the coherent patterns, clusters, and cosmic walls that shape our local corner of the cosmos.

## Presenting "Rotation of the Universe"

Vera's thesis adviser, William Shaw, believed her findings deserved wider attention and encouraged her to present them at the 1950 meeting of the American Astronomical Society (AAS).

Initially, Shaw suggested that he present the work himself. Vera was pregnant with her first child and was not yet a member of the AAS. Presenting a controversial theory before a room full of established astronomers seemed daunting. Vera immediately rejected the suggestion. Years later, she recalled her response: "Immediately, I replied, 'No. I can go.'"

Traveling to the meeting was not easy. Bob and Vera did not own a car, so Vera's parents drove from Washington, D.C., to Ithaca and then continued through the snowy hills of upstate New York to Pennsylvania. Vera later joked that the journey aged her father by twenty years.

At just twenty-two years old and only weeks after giving birth to her first child, Vera stood before a room of professional astronomers to present a paper boldly titled "*Rotation of the Universe.*" The audience was skeptical.

Many of the astronomers in attendance doubted both her conclusions and the limited data on which they were based. Some regarded the work as premature, while others questioned whether such sweeping claims could be justified from observations of only 109 galaxies.

Only one prominent astronomer, Princeton's Martin Schwarzschild, offered significant encouragement. While he did not fully agree with her conclusions, he urged her to gather additional observations and continue investigating the problem.

Both The Astrophysical Journal and The Astronomical Journal ultimately rejected the paper, and they published only a brief abstract of the work. Yet the experience left a lasting impression. Newspapers focused less on the science than on the unusual sight of a young mother presenting original astronomical research. One headline described her as a "*Young Mother Has Own Theory of Universe.*"

Vera later remembered the criticism but refused to be discouraged. Throughout her career, she would encounter skepticism, rejection, and barriers that often had as much to do with her gender as with her science. Rather than retreat, she used these experiences as motivation to continue asking difficult questions.

## Returning to Washington, D.C.

Despite the controversy surrounding her thesis, Vera successfully completed her master's degree in astronomy in 1951. The same year, Bob completed his PhD in mathematics at Cornell, marking a significant milestone for the young couple.

Although many astronomers dismissed her first major research project, the experience taught Vera an important lesson. Scientific progress often begins with observations that challenge established ideas, and new discoveries frequently face resistance before they gain acceptance.

After completing their studies, Vera and Bob returned to Washington, D.C., where Bob accepted a position with the National Bureau of Standards. At just twenty-three years old, Vera found herself balancing two demanding roles: her role as a young mother and as an aspiring astronomer.

For many women of the era, marriage and motherhood marked the end of an academic career. Vera, however, found it impossible to abandon her fascination with the universe. While caring for her infant daughter, she continued reading scientific journals and following the latest developments in astronomy. She later recalled taking her child to the park with copies of *The Astrophysical Journal* tucked under her arm, reading research papers whenever she could find a spare moment.

Before long, she realized that her desire to understand the universe had not diminished. If anything, it had grown stronger. With the encouragement of her husband, Vera decided to continue her education and pursue a doctorate. She enrolled in Georgetown University's astronomy PhD program, one of the few programs in the Washington area that offered evening classes, allowing her to balance family responsibilities with research. The support of both Bob and her parents proved invaluable during these years, as Vera often relied on them while attending classes, conducting research, and raising her young family.

## Doctoral Studies at Georgetown University

At Georgetown University, Vera began studying the properties and motions of galaxies on a much larger scale than she had explored during her master's work. Although she officially enrolled at Georgetown, George Gamow, one of the twentieth century's leading scientific minds, heavily influenced her doctoral research as a theoretical physicist and cosmologist.

Gamow was already famous for his pioneering work in nuclear physics and cosmology and for helping to lay the foundations of what would later become known as the Big Bang theory. He first became aware of Vera's work after her controversial presentation to the AAS in 1950. Although many astronomers dismissed her conclusions, Gamow was intrigued.

Rather than criticizing her ideas, he began asking questions. One question in particular captured Vera's attention: "Is there a scale length in the distribution of galaxies?" The question was deceptively simple. Were galaxies scattered randomly throughout the universe, or was there some larger pattern governing their arrangement? Vera decided to devote her doctoral research to finding the answer.

Ironically, despite becoming her advisor, Gamow did not always embrace modern views regarding women in science. Vera later observed that he did not allow wives to attend some of his lectures. Yet he recognized her talent and encouraged her investigations into the large-scale structure of the universe.

---

**HISTORICAL SPOTLIGHT**

## George Gamow

George Gamow (1904–1968) was a visionary physicist whose ideas helped lay the foundations of modern cosmology. Born in Odessa and later working in the U.S., Gamow combined deep theoretical insight with an irrepressible sense of curiosity. In the 1940s, he and his collaborators Ralph Alpher and Robert Herman developed the first detailed model of the early universe, proposing that it began in a hot, dense state and has been expanding ever since. They predicted the existence of the cosmic microwave background, which is a faint afterglow of the Big Bang. Its detection would not occur until 1965.

Gamow also made major contributions to nuclear physics, explaining how quantum tunneling allows particles to escape atomic nuclei, a key insight into radioactive decay. Known for his playful writing and talent for popular science, he helped bring complex ideas to a wide audience. Today, Gamow is remembered as one of the bold thinkers who transformed the Big Bang from speculation into a scientific theory.

---

## Investigating the Structure of the Universe

At the time Vera began her doctoral studies, most astronomers viewed galaxies as being distributed relatively randomly throughout space. While researchers knew galaxy clusters existed, few actively investigated whether larger patterns might be present. Vera approached the problem as an observer rather than a theorist.

Throughout her career, she would remain fundamentally committed to collecting high-quality observations and allowing the data to guide her conclusions. She was less interested in constructing elaborate theories than in determining what the universe was actually doing.

For her doctoral dissertation, Vera analyzed the distribution of galaxies and searched for evidence of large-scale organization. After examining the available observations, Vera concluded that galaxies did not scatter randomly throughout space. Instead, they tended to cluster together in groups and larger structures. Today, this may seem obvious, but in the early 1950s, it was a controversial conclusion.

Vera's dissertation proposed that galaxies formed vast associations rather than being uniformly distributed throughout the cosmos. In essence, she was identifying the large-scale structure of the universe decades before the subject became a major area of research.

She completed her dissertation, titled On the Nature of the Extragalactic Universe, in 1954. The work showed that galaxies frequently appeared in clusters and groups, suggesting that the universe possessed a far more organized structure than many astronomers had assumed.

## Ahead of Her Time and Earning Her Doctorate

In 1954, at just 26 years old, Vera completed her PhD in astronomy at Georgetown. The accomplishment was remarkable not only for the quality of her research but also for the circumstances under which she achieved it: she was raising a young family during an era when people routinely discouraged women from pursuing scientific careers. Reflecting later, she observed that "women generally required more luck and perseverance than men did," and credited her supportive parents and husband with helping her persist.

Her doctoral research, like her master's work, received little attention when it was first published. Many astronomers regarded her conclusions as speculative, and the limited observational data available in the early 1950s prevented rigorous testing of her ideas. The study of large-scale cosmic structure was still in its infancy, and few recognized the significance of what she had uncovered.

Yet Vera had identified something important. Her dissertation suggested that galaxies did not distribute randomly through space but instead formed larger patterns and groupings. Although later refinements would adjust her specific interpretation, the underlying insight proved remarkably prescient. Over the following decades, increasingly powerful telescopes and extensive galaxy surveys showed that galaxies arrange themselves in clusters, superclusters, filaments, and enormous cosmic walls, forming the intricate network now known as the cosmic web.

More importantly, her doctoral research revealed a characteristic that would define her entire career: Vera consistently noticed patterns in the data that others overlooked. Guided by persistence, intellectual independence, and a willingness to follow the evidence wherever it led, she would eventually make one of the most transformative discoveries in modern astronomy: the first compelling observational evidence for dark matter.

## Georgetown University and Family Life

After completing her doctorate in 1954, Vera entered a new phase of her life. She was now a trained astronomer with original research to her name, but she was also a wife and mother raising a growing family. The challenge of balancing science and family would become a defining feature of her early career.

Following graduation, Vera spent a year teaching mathematics and physics at Montgomery College in Maryland before returning to Georgetown

University. There she began what would become more than a decade of teaching and research, serving as a research associate astronomer, lecturer, and eventually assistant professor of astronomy.

At Georgetown, Vera continued to pursue astronomical research while raising her children. The arrangement was far from easy. Professional astronomy remained overwhelmingly male, and few women attempted to combine an active research career with motherhood. Yet Vera was determined to do both.

## Building a Family

Throughout these years, Vera's greatest supporter was her husband, Bob. Bob, Vera's greatest supporter throughout these years, encouraged her scientific ambitions and helped create a home where her work could flourish at a time when many women were expected to choose between family and career. Together, they raised four children, David, Judy, Karl, and Allan, in a household where family life and scientific curiosity blended naturally.

Balancing motherhood with a growing research career was never easy. Vera later recalled the difficulty of confronting the assumption that she should remain at home while her husband pursued the work he loved. "I realized that as much as we both adored this child," she said, "there was nothing in my background that had led me to expect that he would go off to work each day doing what he loved to do, and I would stay home. I really found it very, very hard."

She worked wherever and whenever she could, often late at night after the children were asleep, spreading charts and papers across the kitchen table once the dishes were done. Colleagues remembered a home filled with both noise and discovery: children, homework, telescope schedules, and research notes all sharing the same space. Her children later joked that they grew up thinking graphs and scientific debates were a normal part of dinner.

By 1963, Vera was recognized as one of the few women of her generation who had managed to sustain a professional scientific career while raising a family largely by herself. Her position at Georgetown offered enough flexibility for her to work from home, allowing her to continue publishing and teaching while caring for her children.

The Rubin household became a remarkable scientific incubator. All four children eventually earned PhDs in the natural sciences or mathematics.

- David (1950), a geologist with the U.S. Geological Survey,
- Judith Young (1952–2014), an astronomer at the University of Massachusetts,
- Karl (1956), a mathematician at the University of California, Irvine,
- Allan (1960), a geologist at Princeton University.

They later credited their mother with making a life of science seem joyful, natural, and attainable. Vera and Bob remained devoted partners for nearly sixty years until Bob died in 2008. Vera's older sister, Ruth Cooper Burg, also forged a distinguished career, becoming an attorney and later an administrative law judge in the U.S. Department of Defense.

## Returning to Research

Although teaching occupied much of her time during the 1950s, Vera never abandoned research. Her doctoral work on the clustering of galaxies had already suggested that the universe was more structured than astronomers had assumed. Now she turned her attention to another question: how galaxies themselves moved. Vera was, above all, an observer.

Unlike many astronomers who focused on theoretical calculations, she preferred gathering precise measurements and letting the data reveal the answer. Her strength lay in designing observational programs, collecting accurate measurements, and identifying patterns that others overlooked.

Throughout the late 1950s and early 1960s, she continued studying the motions of galaxies, steadily building the expertise that would later lead to her most famous discoveries.

## Early Evidence of a Mystery

In 1962, Vera published a paper with six of her students examining the rotation of the Milky Way. The results contained a puzzling observation.

According to existing theories, stars farther from the center of a galaxy should orbit more slowly than those near the center, much as the outer planets of the Solar System move more slowly around the Sun than the inner planets. Instead, Vera's measurements suggested something unexpected.

The rotational speed of the Milky Way appeared to remain relatively constant rather than declining with distance from the center. The rotation curve appeared "flat." At the time, the significance of this finding was not fully appreciated. The paper received little attention, and few astronomers considered it particularly important.

Looking back, however, scientists now recognize it as an early hint of one of the greatest mysteries in modern astronomy. A flat rotation curve implied that more mass existed in the outer regions of the galaxy than visible stars and gas alone could account for. Something unseen appeared to be exerting a gravitational influence. The implications would not become clear until years later, but Vera had already begun to encounter evidence of what would eventually become known as dark matter.

## Observing Galaxies and a New Direction

Vera's research sped up in the early 1960s. Between 1963 and 1964, she collaborated with astronomers Geoffrey and Margaret Burbidge and gained access to the 82-inch telescope at McDonald Observatory in Texas, one of the most powerful instruments of its time. There, she conducted some of her first detailed studies of how stars and gas move within spiral galaxies.

Working systematically, Vera compared the appearance of different spiral galaxies with their internal motions, searching for relationships between a galaxy's structure and the way it rotated. These observations deepened her interest in galactic dynamics and laid the foundation for the discoveries that would later transform astronomy.

By the mid-1960s, however, Vera was balancing more responsibilities than ever. She was teaching at Georgetown, conducting research, publishing scientific papers, and raising four children. The pace was unsustainable. After more than a decade in academia, she made a pivotal decision: to step away from full-time teaching so she could devote herself more fully to research. As she later explained, "Teaching and observing and having four children was just more than I wanted to handle."

That decision marked a turning point in her career. Freed from the demands of a heavy teaching load, Vera entered the most productive period of her scientific life. The observational skills she had developed through years of careful measurement, patient data collection, and a willingness to question accepted ideas would soon lead her to one of the most important discoveries in modern astronomy: compelling evidence that much of the universe consists of invisible dark matter.

## Palomar Observatory

A year before leaving Georgetown to focus entirely on research at the Carnegie Institution, Vera achieved another important milestone in her scientific career.

In 1965, astronomer Allan Sandage invited Vera to conduct observations at the Palomar Observatory in California, home to the famous 200-inch (5-metre) Hale Telescope. Palomar was one of the world's premier astronomical research facilities at the time, and astronomers around the globe highly coveted access to its powerful telescope.

For Vera, the invitation represented far more than an opportunity to gather data. It was also a chance to challenge one of the many barriers that women scientists still faced. Although no formal rule prohibited women from using the telescope, few had received observing time. Vera became the first woman to officially use the Hale Telescope as an observer, a significant breakthrough at a facility long dominated by men.

Her work at Palomar focused on the controversial study of large-scale motions of galaxies, research that would later contribute to her investigations of galaxy dynamics and the structure of the universe. Yet even as she arrived to conduct cutting-edge scientific research, she encountered a reminder of how uncommon women scientists still were. Vera was informed that observing arrangements would be somewhat difficult because the observatory did not have a women's restroom. During her facility tour, she saw that the only restroom said "Men."

Rather than complain, Vera responded with the combination of practicality and quiet humor that would become one of her trademarks. She drew a skirt on a piece of paper, taped it beneath the figure on the restroom sign, and announced, "There you go; now you have a ladies' room."

The story became one of the most famous anecdotes of Vera's career, not because of the bathroom itself, but because it perfectly captured her approach to discrimination. She rarely sought confrontation, yet she refused to accept exclusion as normal. Instead, she met obstacles with determination, humor, and a quiet insistence that women belonged wherever scientific discovery was taking place.

Three decades later, when Vera returned to Palomar in 1995, much had changed. Heating was now in the observing room, the men-only sign had disappeared, and a gender-neutral restroom was available for all astronomers.

**HISTORICAL SPOTLIGHT**

**Palomar Observatory**

Palomar Observatory, located on Palomar Mountain in southern California, is one of the most iconic astronomical sites of the 20th century. Operated by the California Institute of Technology, it is home to the 200-inch Hale Telescope, which was the world's largest effective optical telescope when it began operation in 1949. For decades, Hale's unparalleled light-gathering ability made it the foremost instrument for examining galaxies, quasars, and the universe's large-scale structure.

Palomar became a crucial site for modern cosmology. Its telescopes enabled precise measurements of galaxy distances, redshifts, and rotation curves, work that helped reveal the expanding universe and the presence of dark matter. The observatory also played a historic role in opening doors for women in astronomy; Vera became the first woman officially allowed to observe there in the late 1960s.

Today, Palomar remains an active research center, combining its historic instruments with modern surveys that continue to map and understand the dynamic universe.

## Carnegie Institution's Department of Terrestrial Magnetism

In January 1965, Vera visited her friend Bernard ('Bernie') Burke, an astrophysicist at the Carnegie Institution of Washington's Department of Terrestrial Magnetism (DTM). Never one to shy away from a challenge, Vera walked into Burke's office and directly asked if there was a position available. Vera later joked that Burke looked "as surprised as if she had asked him to marry her."

At the time, the DTM had never employed a woman scientist on its research staff. Nevertheless, Burke arranged for Vera to attend the department's weekly community lunch, where she met the department director, physicist Merle Tuve.

During lunch, Tuve asked Vera to stand at a blackboard and explain her research on galaxy motions. The request was entirely unexpected, but Vera welcomed the opportunity. She outlined her work on galaxy rotation and large-scale galactic motions, demonstrating both her deep understanding of the subject and her ability to interpret complex observational data. The presentation impressed Tuve.

Afterward, he handed Vera one of the department's photographic plates and asked her to analyze the stellar velocity measurements it contained. Vera took the plate back to her Georgetown office, carefully examined the data, and returned her results in short order. Her analysis was both accurate and thorough.

Less than three months later, the DTM offered Vera a position, making her the first woman scientist to join the department's research staff.

**HISTORICAL SPOTLIGHT**

## Carnegie's Department of Terrestrial Magnetism (DTM)

The Department of Terrestrial Magnetism (DTM) was a research division of the Carnegie Institution of Washington, founded in 1904 to study Earth's magnetic field. Over time, its mission expanded dramatically. By the mid-20th century, DTM had become a world-class center for geophysics, seismology, planetary science, and observational astronomy. Its scientists built instruments, developed new techniques, and conducted pioneering research on everything from the structure of Earth's interior to the motions of galaxies. In 2020, DTM merged with Carnegie's astronomy division to form the Earth and Planets Laboratory.

## Meeting Kent Ford

At the DTM, Vera met astronomer and instrument specialist W. Kent Ford, whose technical innovations would prove essential to her future discoveries.

Kent had developed one of the most sensitive spectrographs in the world, incorporating advanced image-tube technology that dramatically increased the ability of telescopes to detect faint objects. The instrument amplified incoming light before it reached the photographic plate, allowing astronomers to obtain detailed spectra from galaxies that had previously been too dim to study effectively. For astronomers interested in galaxy motions, this represented a major technological breakthrough.

Vera immediately recognized its potential. As she later recalled, "Kent was anxious to test the limits of the image tube, and I was anxious to obtain new information about galaxies."

The partnership was a perfect match. Kent possessed extraordinary skill in designing instruments, while Vera excelled at asking important scientific questions and interpreting observational data.

---

**HISTORICAL SPOTLIGHT**

### Kent Ford

Kent Ford (born 1931) is an American astronomer and instrument innovator whose work transformed the capabilities of modern observational astronomy. After earning his PhD at Johns Hopkins University, Ford devoted his early career to developing image-tube spectrographs, instruments that dramatically increased the sensitivity of telescopes. His designs allowed astronomers to measure the motions of extremely faint stars and galaxies, observations that had previously been impossible with photographic plates.

Ford's technical achievements quickly gained recognition. Major observatories across the U.S. adopted his spectrographs, and his instrumentation set new standards for precision in galactic and extragalactic astronomy. He published influential papers on detector technology, contributed to the modernization of optical astronomy, and helped train a generation of scientists in advanced observational methods.

By pushing the limits of what telescopes could detect, Ford opened the door to discoveries that reshaped cosmology. His career stands as a testament to the power of instrumentation: the idea that breakthroughs in science often begin with breakthroughs in the tools we use to explore the universe.

## Measuring Galactic Motion

Working closely with Kent, Vera began using his exceptionally sensitive image-tube spectrograph to investigate how stars and gas moved within spiral galaxies. Kent's instrument captured far more detail than earlier detectors, allowing the pair to study faint, distant regions of galaxies that had previously been inaccessible.

Their work focused on mapping how material circulated around galactic centers. To do this, they collected light from many points across a galaxy and examined the subtle shifts recorded by the spectrograph. Each position along the disk provided a separate measurement, revealing how quickly that region was moving relative to Earth. By assembling these measurements, Vera and Kent could trace how orbital speeds changed from a galaxy's bright inner regions to its dim outer edges.

Night after night, they gathered these velocity data points for numerous spiral galaxies. What they found was unexpected. Instead of slowing down with distance, as classical gravitational models predicted, the outer regions of galaxies were moving far faster than expected. The pattern appeared again and again, regardless of the galaxy's size or brightness.

**SCIENCE SPOTLIGHT**

### Measuring Galactic Motion

Astronomers determine how galaxies rotate by tracking the motions of stars and gas across a galaxy's disk. To do this, they collect light from many different regions of a galaxy and analyze how the light changes as the material moves. Modern spectrographs can record thousands of fine details in a galaxy's spectrum, allowing researchers to extract precise velocity information from even faint outer regions.

To build a rotation profile, astronomers observe a galaxy along its major axis and measure how quickly material on one side moves toward us while material on the opposite side moves away. Each position along the disk provides a separate velocity data point. Plotting these measurements against distance from the center creates a rotation curve, a powerful diagnostic of how mass is distributed throughout the galaxy.

Different tools reveal different parts of a galaxy.

- Optical spectrographs measure the motions of bright stars and glowing gas in the inner regions.
- Radio telescopes, tuned to the 21-centimeter emission of neutral hydrogen, trace motion far beyond the visible edge, where starlight becomes too faint to detect.
- Image tubes and electronic detectors, pioneered in the mid-20th century, dramatically reduced exposure times and made it possible to measure velocities in dozens of galaxies with unprecedented accuracy.

---

**SCIENCE SPOTLIGHT**

## The Doppler Effect

Christian Doppler, the Austrian physicist who first proposed in 1842 that the observed color or pitch of a wave changes when the source and observer move relative to one another, gives his name to the Doppler effect. His idea, initially applied to starlight, became one of the foundational tools of modern astronomy, allowing scientists to measure motion across vast distances.

The Doppler effect describes how the wavelength of a wave changes when the source and the observer move relative to one another. If the source moves toward the observer, the waves compress, and this produces shorter wavelengths and higher frequencies. If the source moves away, it stretches the waves, creating longer wavelengths and lower frequencies.

In astronomy, this shift appears in the spectral lines of stars and galaxies. A shift toward the blue end of the spectrum shows motion toward Earth, while a shift toward the red shows motion away. By measuring these tiny changes in wavelength, astronomers can determine the speeds of stars, galaxies, and gas clouds, making the Doppler effect one of the foundational tools for understanding the structure and motion of the universe.

## Studying Galaxy Rotation

During the early years of their collaboration, Vera and Kent investigated several astronomical problems, including quasars, which had recently been discovered and were attracting significant scientific attention. Vera, however, preferred a question that was less fashionable but potentially more revealing.

She decided to focus on the rotation of galaxies. Specifically, she wanted to understand how stars and gas clouds moved around the centers of spiral galaxies. Astronomers understood how planets orbited the Sun and how stars moved within the Milky Way, but the detailed motions of stars within other galaxies remained poorly understood.

Using Kent's advanced spectrograph, Vera and Kent began measuring the velocities of stars in nearby galaxies, particularly the Andromeda Galaxy (M31), the closest large spiral galaxy to our own. Vera later explained her reasoning "I decided to pick a problem that I could go observing and make headway on, hopefully, a problem that people would be interested in, but not so interested that they would already have been working on it."

## Galaxy Mergers and Counter-Rotating Galaxies

Another area that fascinated Vera was the behavior of galaxies that did not conform to conventional expectations. Through her observations, she discovered examples of galaxies in which some stars and gas rotated in the opposite direction from the rest of the galaxy. This phenomenon, known as counter-rotation, challenged the prevailing assumption that all material within a galaxy should move in the same direction.

Such unusual motions suggested that galaxies had far more complex histories than astronomers had previously believed. Vera proposed that these systems were evidence of past interactions between galaxies. When galaxies pass close to one another or collide, gravity can dramatically reshape them, pulling streams of stars and gas into long tidal tails, triggering bursts of star formation, and sometimes causing two galaxies to merge into a single larger system.

These observations provided important evidence that galaxy mergers are a fundamental part of galactic evolution. Rather than being isolated and unchanging objects, galaxies were increasingly understood as dynamic systems that grow and evolve through repeated interactions over billions of years.

## The Rubin-Ford Effect

In the early 1970s, Vera renewed her long-standing interest in the large-scale motions of galaxies. Working with Kent and his highly sensitive spectrograph, she began measuring the velocities of distant galaxies with a precision that had previously been impossible.

Vera's earlier research at Cornell had suggested that galaxies might not simply be moving apart because of the expansion of the universe. Instead, they appeared to exhibit additional motions that the standard Hubble flow alone could not explain. Two decades later, improved instruments finally allowed her to investigate the question more thoroughly.

Studying nearby spiral galaxies, particularly the Andromeda Galaxy (M31), Vera and Kent discovered evidence that large groups of galaxies appeared to be moving together through space. Rather than spreading out evenly in all directions as expected from the expansion of the universe, these galaxies seemed to share a common motion.

This phenomenon became known as the Rubin-Ford Effect. The finding was controversial at the time because it suggested that galaxies across vast regions of space, stretching over tens of millions of light-years, could experience gravitational influences on a much larger scale than astronomers had previously conceived.

Although later observations refined the original interpretation, Vera's work helped establish the existence of what astronomers now call large-scale streaming motions, cosmic flows in which entire groups and clusters of galaxies move together through space under the influence of gravity. The research provided important clues about the large-scale structure of the universe and showed Vera's ability to identify patterns in astronomical data that others had overlooked.

---

**SCIENCE SPOTLIGHT**

### Newtonian Physics

Newtonian physics refers to the system of laws and principles developed by Isaac Newton in the late seventeenth century to describe the motion of objects and the forces acting upon them. At its core are Newton's three laws of motion and his law of universal gravitation, which together explain how bodies move, accelerate, and interact across space. These ideas provided the first coherent mathematical framework for predicting planetary orbits, falling objects, tides, and the behavior of everyday physical systems.

For over two centuries, Newtonian physics served as the foundation of classical mechanics and remains remarkably accurate for most motions at human scales, slow speeds, moderate energies, and weak gravitational fields. Even today, astronomers use Newton's laws to model the rotation of galaxies, the dynamics of star clusters, and the gravitational signatures that first hinted at the presence of dark matter.

---

## The Galaxy Rotation Problem

Vera's most influential work grew from a deceptively simple question: How do galaxies rotate? Her interest began in 1962, when she and six Georgetown graduate students examined the Milky Way's rotation and noticed that stars far from the center seemed to move faster than expected. According to Newtonian gravity, outer stars should orbit more slowly, just as planets far from the Sun move more slowly than those close in. The early hints were intriguing but inconclusive.

By the late 1960s, Vera had turned almost entirely to this puzzle. Working with Kent, she used his exceptionally sensitive spectrograph to measure stellar and gas motions in spiral galaxies. They spent long nights at Lowell Observatory and Kitt Peak, gathering precise velocity data. Vera later noted that Kent's device reduced exposure times dramatically, making possible a new era of galactic measurement.

Using Kent's spectrograph, they mapped how fast stars moved at increasing distances from galactic centers. They expected rotation speeds to drop with distance. Instead, on a clear night at Kitt Peak while observing Andromeda, they saw something astonishing: stars in the galaxy's outskirts were moving almost as fast as those near the center. The rotation curve was flat, not declining. "The surprises came very quickly," Vera wrote. "By the end of the first night, we were puzzled by the shape of the rotation curve."

If galaxies contained only the visible stars, gas, and dust, these fast-moving outer stars should have drifted away into space. Yet they remained bound. As Vera and Kent expanded their survey, they found the same pattern in galaxy after galaxy. Flat rotation curves were not an anomaly; they were a universal feature of spiral galaxies.

In 1970, Vera, Kent, and colleagues published their landmark study of Andromeda in *The Astrophysical Journal*. The study offered some of the strongest observational evidence that galaxies contain far more mass than can be seen. Vera remained cautious, gathering years of additional data before drawing broad conclusions. But as the evidence accumulated, the mystery deepened: something unseen was shaping the motion of every spiral galaxy.

The galaxy rotation problem had become impossible to ignore.

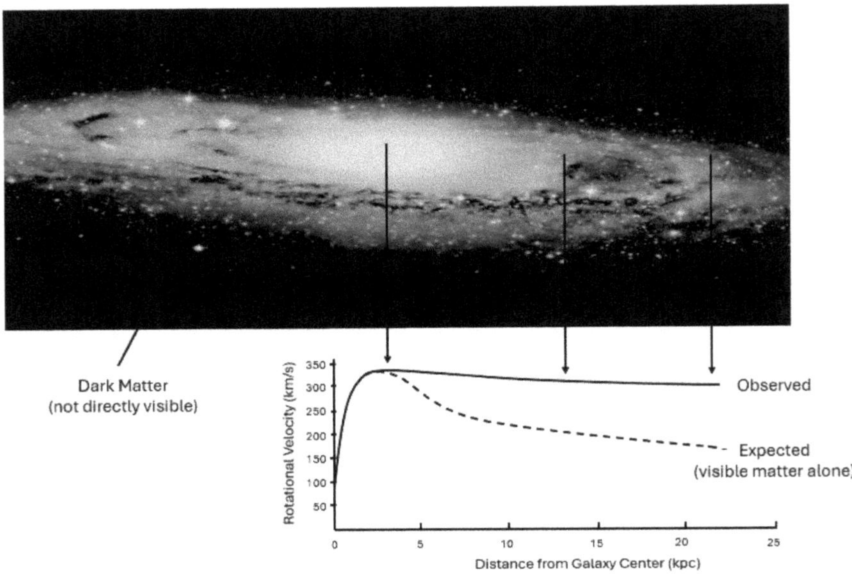


Figure 4. Simplified illustration of a galaxy rotation curve. The unexpectedly high orbital speeds of stars far from the centers of galaxies provided some of the strongest observational evidence for the existence of dark matter. Original illustration by the author, inspired by the pioneering work of Vera Rubin and Kent Ford.

# Evidence of Dark Matter

As Vera expanded her observations through the 1970s, a striking pattern emerged across the universe. Every spiral galaxy she measured displayed the same unexpected behavior: outer stars were orbiting far too quickly to be held in place by the gravity of visible matter alone. What began as a puzzling observation in the Andromeda Galaxy soon proved to be a universal feature of spiral galaxies.

Using Kent's advanced spectrograph, Vera and Kent produced rotation curves for dozens of galaxies. Regardless of size, brightness, or structure, the curves remained flat, a clear indication that the visible stars represented only a small fraction of each galaxy's total mass.

Independent teams soon confirmed the pattern. Radio astronomers mapping hydrogen gas in the farthest reaches of galaxies found the same high orbital speeds. Observations of lenticular and irregular galaxies reinforced the conclusion: the luminous matter was insufficient to explain galactic motion.

Vera's results revived an idea first raised in 1933 by Fritz Zwicky, who argued that galaxy clusters contained far more mass than their visible components suggested. A lack of supporting evidence led critics to dismiss his claim. Vera's galaxy-scale observations provided the missing foundation.

Meanwhile, theorists such as Jeremiah Ostriker, James Peebles, and Amos Yahil were running computer simulations of galaxy evolution. Their models showed that galaxies composed only of visible matter would be unstable; they would distort, collapse, or form bars. But when the simulations included massive, extended halos of unseen material, galaxies behaved realistically. Theory and observation led to the same conclusion: enormous halos of invisible mass must embed galaxies.

By the early 1980s, evidence from optical measurements, radio studies, galaxy cluster dynamics, and gravitational lensing made the case overwhelming. Astronomers accepted that most of a galaxy's mass is invisible. Today, astronomers estimate that this substance, dark matter, makes up about 85% of all matter in the universe.

Although its true nature remains unknown, Vera's meticulous observations transformed dark matter from a speculative idea into a central framework of modern cosmology. She revealed that the universe is shaped not primarily by the stars we see, but by a vast, unseen structure that governs galaxies, clusters, and the cosmic web itself.

**SCIENCE SPOTLIGHT**

### Dark Matter

Dark matter is the name astronomers give to the invisible mass that shapes the universe. It does not emit, absorb, or reflect light, which means no telescope can see it directly. Gravity reveals its presence by how it pulls on stars, bends light, and holds galaxies together.

The strongest clues come from motion. Stars in the outer regions of galaxies orbit far faster than the visible matter alone can explain. Galaxy clusters behave as though they contain far more mass than we can see, and distant starlight curves around unseen structures in a phenomenon called gravitational lensing. Across every scale, the same pattern appears: there is more gravity than visible matter can provide.

Although its exact nature remains unknown, dark matter acts as the hidden framework of the cosmos. It forms vast halos around galaxies, anchors galaxy clusters, and shapes the large-scale web of structure that fills the universe. Ordinary matter, stars, planets, gas, and everything we can see account for only a small fraction of the total. The rest is this mysterious, invisible substance whose gravity quietly sculpts the universe.

## Scientific Leadership and Recognition

Vera and Kents's discoveries regarding galaxy rotation and dark matter did more than transform astronomy; they opened entirely new areas of research that continue to shape astrophysics and particle physics today. What began as an attempt to understand the motions of galaxies evolved into one of the most important scientific investigations of the twentieth century.

As evidence for dark matter accumulated throughout the 1970s and early 1980s, Vera became one of the most respected voices in astronomy. Her careful observations, meticulous analysis, and cautious interpretation of results earned the trust of the scientific community. Unlike many researchers, Vera rarely made bold claims beyond what her data could support. Instead, she allowed the evidence to speak for itself.

Alongside her own research, Vera took on important leadership roles within the scientific community. From 1972 to 1977, she served as Associate Editor of *The Astronomical Journal*, helping to evaluate and oversee the publication of new astronomical research. She then served as Associate Editor of *Astrophysical Journal Letters* from 1977 to 1982, one of the most influential journals in astronomy. In these positions, Vera played an important role in maintaining scientific standards and supporting the communication of new discoveries.

Although Vera was becoming increasingly famous within scientific circles, she remained committed to making astronomy accessible to the public. Throughout her career, she gave lectures, took part in public outreach, and encouraged scientific literacy. She believed that the wonders of the universe belonged to everyone, not just professional scientists.

Her ability to explain complex ideas clearly made her a popular speaker. In 1997, she published *Bright Galaxies, Dark Matter*, a collection of essays and articles written for a general audience. The book offered readers insight into both the science behind her discoveries and the excitement of astronomical research.

By the mid-1980s, Vera's work had gained worldwide recognition. At meetings of the International Astronomical Union (IAU), she presented the growing body of evidence supporting dark matter and challenged astronomers to rethink their understanding of the universe.

Reflecting on how much remained unknown, Vera famously remarked, "In a spiral galaxy, the ratio of dark-to-light matter is about a factor of 10. That's probably a good number for the ratio of our ignorance to knowledge. We're out of kindergarten, but only in about third grade."

## Scholar, Mentor, Pioneer

Throughout her life, Vera devoted as much energy to supporting people as she did to studying galaxies. She believed that science should be open to anyone with the curiosity and determination to pursue it, and she worked tirelessly to make that ideal a reality.

Vera's warmth, fairness, and generosity became legendary among colleagues and students. Deidre Hunter, a longtime collaborator, described her as determined, stubborn in the best way, and unwaveringly just, qualities that helped her navigate and reshape a field that had long excluded women. Vera never forgot the barriers she faced as a young scientist: discouragement from teachers, rejection from graduate programs, and years of being the only woman in the room. Those experiences fueled her commitment to ensuring that others would not face the same obstacles.

Mentorship became one of her defining contributions. Vera encouraged young scientists, especially women, to apply for faculty positions, present at conferences, and pursue leadership roles. She offered advice at all hours, famously saying she was "available twenty-four hours a day to women astronomers." Many who later became leaders in the field credited her with giving them the confidence to continue. Wendy Freedman wrote to her in 1986, "If things are any easier for women like me, it's because women like you have taken the time and energy to help make it so."

Vera also pushed for institutional change. She challenged conferences with few or no female speakers, urged hiring committees to consider qualified women, and advocated for broader representation on review panels and scientific boards. In 1973, she co-authored an influential report on the status of women in astronomy for the AAS, helping lay the groundwork for what became the AAS Committee on the Status of Women in Astronomy.

When the National Academy of Sciences elected her in 1981, making her only the second female astronomer ever admitted, Vera immediately began pressing the Academy to elect more women. Her election represented not only recognition of her scientific accomplishments but also a significant milestone for women in astronomy, a field that had long excluded them from many of its highest ranks.

Her advocacy extended beyond gender. Vera believed that science thrived when people from diverse backgrounds welcomed and supported others. She challenged exclusionary practices at institutions such as the Cosmos Club and used her appointment to the Pontifical Academy of Sciences to highlight the contributions of women and other underrepresented groups in science.

Vera's philosophy was simple and memorable. In *Bright Galaxies, Dark Matter*, she wrote, "There is no problem in science that can be solved by a man that cannot be solved by a woman." She often reminded audiences that "half of all brains are in women," a reminder that talent is universal even when opportunity is not.

Alongside her advocacy, Vera remained committed to public education. Through lectures, interviews, and writing, she helped make astronomy accessible to general audiences, sharing the wonder of the universe with people far beyond the scientific community.

By the end of her career, Vera had become far more than a pioneering astronomer. She was a mentor, a role model, and a force for cultural change within science. Her influence, like the dark matter she helped reveal, continues to shape astronomy in ways that are not always visible but remain profoundly important. Through her discoveries, her mentorship, and her insistence on fairness, Vera transformed both our understanding of the universe and the opportunities available to those who seek to explore it.

## Later Research and Legacy

Although Vera became famous for providing the strongest observational evidence for dark matter, her scientific work did not end there. Throughout the 1980s, 1990s, and early twenty-first century, she remained an active and productive researcher, continuing to investigate the structure and evolution of galaxies.

One area that particularly interested Vera was the phenomenon of counter-rotation in galaxies. She discovered systems in which some stars and gas rotated in one direction while other components rotated in the opposite direction. These unusual motions suggested galaxies could experience complex interactions, collisions, and mergers over their lifetimes.

Such observations helped astronomers better understand how galaxies grow and evolve. Vera proposed that some of these strange systems may have formed when two galaxies moving in different directions merged together. She also studied the ways in which gravitational interactions between neighboring galaxies could distort their shapes and influence their motions.

Her work contributed to a clearer understanding that galaxies are not isolated objects but continually evolve through their interactions with the wider universe.

Vera remained scientifically active for decades after her dark matter discoveries. She authored over 200 scientific papers and continued publishing research well into her eighties. Her work influenced generations of astronomers and helped establish many of the foundations of modern observational cosmology.

Besides her research, Vera remained deeply involved in scientific education and public outreach. She wrote articles and books aimed at general audiences and worked to make astronomy accessible to people outside the scientific community. Her collection, *Bright Galaxies, Dark Matter,* introduced many readers to the excitement of astronomical discovery.

Recognition for her contributions continued to grow throughout her career. Yet despite repeated nominations and widespread support from colleagues, Vera never received the Nobel Prize, a fact many astronomers continue to regard as one of the most significant omissions in the award's history.

## Honors and Recognition

As evidence for dark matter continued to accumulate and astronomers increasingly accepted its existence, Vera's reputation grew around the world.

What began as a controversial observation eventually became one of the most important discoveries in modern astronomy. Over the course of her career, Vera received numerous awards and honors recognizing both her scientific achievements and her contributions to the astronomical community.

Throughout the following decades, honors continued to accumulate. In 1993, President Bill Clinton awarded Vera the National Medal of Science, the highest scientific honor given by the United States government. Her citation praised her "significant contributions to the realization that the universe is more complex and mysterious than had been imagined." The award acknowledged the profound impact her observations had made on humanity's understanding of the cosmos.

Further recognition followed. In 1994, Vera received the Dickson Prize in Science from Carnegie Mellon University and the Henry Norris Russell Lectureship from the American Astronomical Society, one of the society's highest distinctions.

Perhaps one of her most significant international honors came in 1996, when she became only the second woman ever to receive the Gold Medal of the Royal Astronomical Society in London. The first female recipient was the astronomer Caroline Herschel, who received the medal in 1828. The nearly 170-year gap between the two awards highlights how rare such recognition for women in astronomy remained.

That same year, President Clinton appointed Vera to the National Science Board, the governing body of the National Science Foundation. Through this role, she helped shape science policy and advocate for scientific research at a national level.

In 2004, the National Academy of Sciences awarded Vera the James Craig Watson Medal in recognition of her pioneering work on dark matter and her extraordinary mentorship of young scientists. The citation praised both her "seminal observations of dark matter in galaxies" and her generous support of aspiring astronomers.

Over her lifetime, Vera received more than a dozen major scientific awards, four honorary doctorates, and countless other honors. *Discover* magazine named her one of the fifty most important women in science in 2002. In 2012, the American Physical Society designated the Carnegie Institution's Broad Branch Road campus a historic site in recognition of her groundbreaking dark matter research.

To honor Vera's discovery of dark matter and her role in broadening scientific participation, the American Women Quarters Program featured her in 2025. The recognition reflected both her scientific achievements and her tireless efforts to create opportunities for future generations of scientists.

## Rubin Papers and Personal Archive

In 2001, Vera donated her personal papers to the Library of Congress. The collection provides an extraordinary record of her life and career and offers historians a rare opportunity to understand both the scientist and the person behind the discoveries.

One thing quickly became apparent to researchers examining the archive: Vera rarely threw anything away. Her filing cabinets contained decades of correspondence, notes, drafts, research materials, photographs, and personal reflections. Together, these documents provide an unusually detailed picture of her scientific career and personal life.

The collection also revealed an important lesson for historians. Comparing Vera's contemporary letters and notes with interviews she gave later in life showed that memories can evolve over time. Like many people reflecting on events decades later, Vera sometimes remembered experiences differently than she had recorded them. The archive showed the value of combining oral histories with original documents when reconstructing a person's life.

Among the most surprising discoveries within the collection was a large body of correspondence between Vera and Father Martin McCarthy, a Jesuit astronomer at the Vatican Observatory. Their friendship began when McCarthy visited Georgetown and continued for over 30 years.

The surviving letters reveal a relationship that extended well beyond professional astronomy. McCarthy and Vera exchanged thoughts on scientific research, family life, travel, world events, religion, and personal challenges. Early in her career, Vera appears to have regarded McCarthy as a mentor. Over time, however, their relationship evolved into a close and trusted friendship.

Vera's responses have mostly disappeared, even though McCarthy's letters are still around. Nevertheless, the correspondence provides a rare glimpse into the personal side of a scientist who was often intensely private. Historians have suggested that few people outside her immediate family understood Vera as well as McCarthy did.

The archive continues to serve as a valuable resource for scholars studying both the history of astronomy and the experiences of women in science during the twentieth century.

---

**HISTORICAL SPOTLIGHT**

## Father Martin McCarthy, S.J.

Father Martin McCarthy, S.J. (1917–2010) was a Jesuit astronomer whose career at the Vatican Observatory helped shape mid-20th-century stellar astronomy. Trained in both physics and theology, McCarthy joined the Observatory in the 1950s, where he became known for his meticulous work on stellar classification, variable stars, and the structure of the Milky Way. His research contributed to the Observatory's longstanding mission of advancing scientific knowledge as a form of intellectual inquiry grounded in faith.

McCarthy played a key role in modernizing the Vatican Observatory's research programs, strengthening collaborations with universities and observatories around the world. He also served as an educator and mentor, helping train young astronomers and promoting the idea that scientific discovery and religious vocation could coexist harmoniously. His career exemplified the Jesuit tradition of scholarship: rigorous, curious, and dedicated to understanding the universe in all its complexity.

---

## Death

On December 25, 2016, Vera died in Princeton, New Jersey, at the age of 88. She had been living in an assisted living facility and was suffering from complications associated with dementia.

News of her death prompted tributes from around the world. Colleagues, former students, and scientific organizations remembered her not only as one of the most important astronomers of the twentieth century but also as a generous mentor and advocate for equality in science. The Carnegie Institution, where she had conducted much of her groundbreaking research, described her as a national treasure.

Looking back on her life, Vera once wrote, "My life has been an interesting voyage. I became an astronomer because I could not imagine living on Earth and not trying to understand how the Universe works."

It was a fitting reflection from a scientist whose curiosity never faded. Through her discoveries, Vera changed our understanding of the cosmos. Through her mentorship and advocacy, she changed the lives of countless scientists who followed in her footsteps. Her influence continues to be felt throughout astronomy, ensuring that her legacy will endure for generations to come.

By the time of her death in 2016, Vera had transformed our understanding of the universe. Her observations provided the strongest evidence that most of the matter in the cosmos is invisible, a discovery that fundamentally changed modern astronomy. Yet her legacy extends far beyond dark matter. Through her mentorship, advocacy, and determination, she opened doors for countless women in science and helped create a more inclusive scientific community.

Today, astronomers continue searching for the mysterious dark matter that Vera helped reveal. Although it remains unseen, its influence can be measured throughout the cosmos. In many ways, the same can be said of Vera herself. Her discoveries, her advocacy, and the opportunities she created for others continue to shape astronomy long after her lifetime.

## Nobel Prize Controversy

Despite receiving many of the highest honors available to a scientist, one award always seemed conspicuously absent from Vera's list of achievements: the Nobel Prize.

For many years, Vera was considered a leading candidate for the award. Her work provided some of the strongest observational evidence for dark matter, a discovery that fundamentally changed astronomy and cosmology. Yet no Nobel Prize ever came.

The omission has remained a source of debate among scientists and historians. Some have argued that failing to recognize Vera reflected the broader challenges women faced in receiving credit for major scientific discoveries. Others believe the oversight resulted from the Nobel Committee's difficulty in identifying a specific discovery related to dark matter that fits the prize's criteria. Still others have described the omission simply as one of the Nobel Prize's most significant missed opportunities.

Astrophysicist Emily Levesque summarized a common view among astronomers when she remarked, "She has won every scientific award that a scientist can win except the Nobel Prize. At this point, it seems more like a glaring omission on the Nobel Prize committee's part."

The debate intensified following Vera's death. Since Nobel Prizes are not awarded posthumously, the possibility of future recognition ended. Since then, discussions have frequently included her name among those scientists whose contributions were widely regarded as Nobel-worthy but never formally recognized.

Even without a Nobel Prize, Vera undeniably impacted science. Her observations transformed our understanding of galaxies and revealed that most of the matter in the universe is invisible. Few discoveries have altered humanity's view of the cosmos so profoundly.

Today, astronomers continue searching for the true nature of dark matter, building on foundations that Vera helped establish decades ago. While the Nobel Prize may never bear her name, her scientific legacy remains secure.

Few astronomers have changed our understanding of the universe as fundamentally as Vera.

## Legacy

Vera's legacy extends far beyond her groundbreaking discovery of evidence for dark matter. Through her scientific achievements, mentorship, and advocacy for women in science, she transformed both our understanding of the universe and the opportunities available to those seeking to explore it.

Vera's observations of galaxy rotation fundamentally changed modern astronomy. Before her work, astronomers believed that most of the mass in the universe was visible through stars, planets, gas, and dust. Her research showed that galaxies held significantly more matter than was visible. Today, astronomers estimate that ordinary matter makes up only a small fraction of the universe, while dark matter and dark energy account for most of its contents. Over fifty years after Vera's first observations, scientists around the world continue searching for the true nature of dark matter, making it one of the greatest unsolved mysteries in science.

Her influence is obvious throughout modern astronomy. Major research facilities around the world, such as the European particle physics laboratory CERN, routinely test theories about dark matter. Entire fields of astrophysics and cosmology have developed from questions raised by Vera's observations. What began as a puzzling discrepancy in galaxy rotation curves evolved into one of the most important scientific discoveries of the twentieth century.

However, Vera's influence extended beyond scientific research. Throughout her career, she worked tirelessly to support young scientists, particularly women entering astronomy. She encouraged students, mentored early-career researchers, and advocated for greater opportunities for women in science. Many of today's leading astronomers credit Vera's guidance and encouragement as instrumental in their careers.

Her accomplishments were particularly remarkable given the obstacles she faced. Vera built a distinguished scientific career while raising four children during a period when professional astronomy remained overwhelmingly male-dominated. Despite being denied opportunities available to her male colleagues and repeatedly confronting discrimination, she persisted and ultimately became one of the most respected astronomers of her generation.

Following her death, many institutions and organizations moved quickly to honor her contributions. The AAS's Division on Dynamical Astronomy renamed its annual early-career award to the Vera Rubin Early Career Prize, recognizing both her scientific achievements and her commitment to supporting junior researchers. The organization noted that the honor was especially fitting because Vera was renowned not only for her research but also for her kindness, generosity, and encouragement of future generations of scientists.

Today, Vera's name appears throughout the scientific world. In 2017, a ridge on Mars was named the Vera Rubin Ridge in her honor. Her name also graces asteroid 5726 Rubin, satellites, scientific awards, educational programs, and major astronomical facilities. The Carnegie Institution, where she spent much of her career, maintains extensive archives and photographic collections documenting her life and work.

One of the most significant tributes came with the naming of the world's newest major astronomical facility, the Vera C. Rubin Observatory. Designed to conduct one of the most ambitious sky surveys ever attempted, the observatory reflects Vera's lifelong dedication to exploring the unknown regions of the universe. Its mission to uncover hidden objects and phenomena makes it an appropriate tribute to the scientist who revealed that most of the universe is invisible.

Recognition of Vera's influence has also extended beyond science. The last episode of the television series *Cosmos: A Spacetime Odyssey* featured her, introducing her story to a new generation. The U.S. honored her on currency in 2025 with the release of the Dr. Vera Rubin quarter as part of the American Women Quarters Program. In 2024, technology company NVIDIA announced that a future generation of its data-center computing architecture would bear the names Vera and Rubin in tribute to her scientific contributions.

Through her books, scientific papers, lectures, and mentorship, Vera hoped future generations would continue exploring the mysteries she helped uncover. Reflecting on humanity's growing understanding of the cosmos, she once wrote, "We have peered into a new world and have seen that it is more mysterious and more complex than we had imagined. Still, more mysteries of the universe remain hidden. Their discovery awaits the adventurous scientists of the future. I like it this way."

Those words capture the spirit that defined her career. Vera did not simply answer questions about the universe; she revealed entirely new questions to ask. Her discoveries opened a door onto a hidden cosmos, inspiring generations of scientists to continue the search for answers. Today, every investigation into dark matter, every study of galaxy formation, and every young astronomer inspired by her story forms part of her enduring legacy.

## Vera C. Rubin Observatory Designation Act

One of the most meaningful honors given to Vera came several years after her death. On January 6, 2020, at the 235th meeting of the American Astronomical Society in Honolulu, astronomers received official confirmation that the Vera C. Rubin Observatory Designation Act had become United States law. The legislation, signed on December 20, 2019, formally decreed that the major new astronomical facility under construction on Cerro Pachón in Chile would bear Vera's name.

The recognition was historic: no one had ever named a major U.S. national observatory after a woman. Formerly known as the Large Synoptic Survey Telescope (LSST), it was renamed to honor Vera's pioneering contributions to dark matter research and her lifelong advocacy for women in science. Congress explicitly acknowledged both aspects of her legacy, noting that her observations provided some of the strongest evidence for dark matter and that her mentorship opened doors for generations of scientists.

Paul Dabbar, then Under Secretary for Science at the U.S. Department of Energy, captured the significance of the tribute: "Dr. Rubin's life and singular achievements as a scientist remain a model for all those seeking to satisfy humanity's unceasing curiosity about our universe."

## The Rubin Observatory and the Dark Matter Legacy

One of the observatory's earliest working names was the "Dark Matter Telescope," reflecting the very questions Vera helped bring to the forefront of modern astronomy. Today, the NSF–DOE Vera C. Rubin Observatory stands as a continuation of that scientific journey.

The facility houses an innovative 8.4-meter telescope and the largest digital camera ever built for astronomy, a 3.2-gigapixel instrument capable of capturing extraordinarily detailed images of the night sky. Over its ten-year mission, the Legacy Survey of Space and Time (LSST) will repeatedly image the southern sky, creating an unprecedented time-lapse record of the universe. It is expected to collect more data than all previous optical observatories combined in its first year alone.

The observatory was designed to address four major scientific goals:

1. Probing dark matter and dark energy through gravitational lensing, galaxy clustering, and cosmic expansion.
2. Cataloging asteroids, comets, and other Solar System bodies, improving planetary defense, and revealing the structure of the outer Solar System.
3. Mapping the Milky Way to reconstruct its formation and evolution.
4. Capturing transient events: supernovae, variable stars, black hole outbursts, and other rapidly changing phenomena.

Even during early commissioning, the observatory showed its extraordinary potential. Within weeks of test operations, astronomers identified over 11,000 previously unknown asteroids and refined the orbits of over 80,000 known objects, including dozens of near-Earth objects and hundreds of distant trans-Neptunian worlds. Over the course of the full survey, scientists expect it to discover millions of new Solar System bodies and dramatically improve our understanding of the region beyond Neptune, perhaps even shedding light on the hypothetical Planet Nine.

Yet the observatory's most profound impact may lie in its ability to illuminate the invisible universe that fascinated Vera throughout her career. By mapping billions of galaxies and tracing the cosmic web, it will help scientists investigate the dark matter that binds galaxies together and the dark energy driving the universe's expansion.

More than any award or monument, the Vera C. Rubin Observatory embodies the spirit of Vera's life's work. Each night it scans the sky, searching for hidden structures and unseen forces, carrying forward the questions she devoted her career to answering. Just as Vera revealed that most of the universe is invisible, the observatory that bears her name is helping humanity uncover what still lies beyond the reach of human sight.

## Why She Matters Today

Vera's legacy continues to shape both the science of the universe and the culture of the people who study it. Vera's meticulous measurements of galaxy rotation showed that dark matter constitutes most of the cosmos. The presence of this unseen substance remains a central pillar of modern astrophysics. Every major cosmological model used today, from simulations of galaxy formation to studies of the universe's expansion, depends on the phenomenon she helped bring into focus. Vera didn't just change a theory; she changed the framework through which scientists understand the structure and evolution of the universe.

But her influence reaches far beyond the stars. Vera transformed the culture of science itself. When women were routinely excluded from observatories, conferences, and faculty positions, she quietly but firmly insisted that talent deserved opportunity. She challenged institutions to do better, advocated for women at every career stage, and modeled a version of scientific life that included both intellectual ambition and a rich family world. Her example continues to resonate with researchers who are navigating the same tensions between work, identity, and belonging.

Her impact is also visible in the next generation. Vera's insistence that science should be open to everyone helped shift expectations for equity and representation in astronomy. Many of today's women astronomers cite her as the person who made their careers feel possible. Her life remains a reminder that scientific progress is not only about data and discovery but also about who gets to take part in the search for knowledge.

The observatory that now bears her name, the Vera C. Rubin Observatory, carries her legacy into the future. Each night it scans the sky for the unseen, mapping billions of galaxies and probing the dark matter and dark energy that define the universe. Its mission echoes Vera's own: to reveal what lies beyond the limits of human sight and understanding. In this way, her influence is not only historical but ongoing. The questions she opened continue to guide the most ambitious scientific project of our time.

Vera matters today because she expanded our view of the universe and widened the doors of the scientific community. She showed that discovery and compassion, rigor and fairness, intellect and humanity can coexist, and that when they do, science becomes not only more accurate, but more just.

## Timeline

| Year | Event |
|---|---|
| 1928 | Born in Philadelphia, U.S. |
| 1948 | Graduated from Vassar College; married Robert Rubin |
| 1950 | Presentation at AAS; David born |
| 1951 | Awarded Master's degree at Cornell University |
| 1952 | Judith born |
| 1954 | Awarded PhD from Georgetown University |
| 1956 | Karl born |

| | |
|---|---|
| 1960 | Allan born |
| 1962 | Paper published on the rotation of the Milky Way |
| 1965 | Joined DTM; Met Kent Ford; Worked at Palomar Observatory |
| 1970 | Published landmark study of the Andromeda Galaxy |
| 1972-82 | Served as Associate Editor of the *Astronomical Journal (1972-77)* and *Astrophysical Journal Letters (1977-1982)* |
| 1997 | Published *Bright Galaxies, Dark Matter* |
| 1980s – 90s | Worldwide recognition; elected to the National Academy of Sciences; awarded National Medal of Science; received Gold Medal of the Royal Astronomical Society |
| 2001 | Donated personal papers to the Library of Congress |
| 2014 | Retired |
| 2016 | Dies at age 88 |

"Science is competitive, aggressive, demanding. It is also imaginative, inspiring, uplifting. You can do it, too. … Each one of you can change the world, for you are made of star stuff, and you are connected to the universe."

— *Vera Rubin*

# CONCLUSION

The women featured in this book lived in different centuries, worked in different scientific fields, and followed very different paths. Yet their stories reveal remarkable similarities.

Each possessed an intense curiosity about the world around them. Each faced obstacles that threatened to limit their opportunities. And each persevered long enough to make discoveries that changed how we understand ourselves, our planet, and our universe.

Ada Lovelace imagined the possibilities of computing more than a century before modern computers existed. Marie Curie uncovered the hidden power of radioactivity and became the first person to win Nobel Prizes in two different sciences. Jane Goodall transformed the study of animal behavior through patient observation and challenged assumptions about what makes humans unique. Rosalind Franklin's precise X-ray diffraction research helped reveal the structure of DNA, one of the most important discoveries in biology. Vera Rubin provided compelling evidence that much of the universe consists of invisible matter that scientists are still trying to understand today.

Their achievements remind us that science is not simply a collection of facts. It is a process of exploration. Every discovery raises new questions. Every answer reveals new mysteries.

Many of the questions these women helped uncover remain active areas of research today. Scientists continue to investigate the nature of dark matter, study the genetic code, develop new technologies inspired by early computing concepts, and work to better understand animal behavior and ecosystems. Their discoveries were not the end of the story; they were the beginning of new chapters.

The world these women entered was often reluctant to accept them. Some institutions denied them educational opportunities. Some struggled to get funding, recognition, or professional positions. Others saw their contributions overlooked or minimized. Yet their work ultimately spoke for itself.

Today, opportunities for women in science are greater than they were during the lifetimes of these pioneers. Still, their stories remain important reminders that talent and curiosity can come from anywhere. Progress is often driven by people willing to challenge accepted ideas, ask difficult questions, and persist when others doubt them. Perhaps the greatest lesson these women offer is that science belongs to everyone.

The next great discovery may come from a laboratory, a classroom, an observatory, a field station, or a home workshop. It may come from someone who is told that their ambitions are unrealistic, their questions are unimportant, or their goals are impossible. History suggests otherwise.

The women in this book proved that determination, curiosity, and careful observation can change the world. Their discoveries continue to shape our lives, and their stories remind us that countless mysteries still await exploration.

Perhaps Vera Rubin expressed this idea best when she said:

"We have peered into a new world and have seen that it is more mysterious and more complex than we had imagined. Still, more mysteries remain hidden. Their discovery awaits the adventurous scientists of the future."

And some of those future scientists may already be reading this book.

# REFERENCES

## Chapter 1: Marie Curie


*24 rue de la Glacière: the historic address of the Curies in Paris 13th.* (n.d.). Century 21 Arago - the Gobelins. Retrieved June 24, 2026, from https://www.lesgobelins-immobilier.fr/fr/blog/24-rue-de-la-glaciere-l-adresse-ou-pierre-et-marie-curie-ont-decouvert-le

*April 19, 1906 - Death of Pierre Curie, Nobel Prize winner for discovering radioactivity and the chemical elements radium and polonium - Rincón educativo.* (2024, December 20). Rincón Educativo. https://rinconeducativo.org/en/anniversaries/april-19-1906-death-pierre-curie-nobel-prize-discovering-radioactivity-radium-polonium/

Association Curie et Joliot-Curie. (2026, February 9). *Joliot-Curie Family Timeline | Association Curie et Joliot-Curie.* Association Curie Et Joliot-Curie. https://curie-joliotcurie.fr/en/about-us/joliot-curie-family-timeline/#section-943-1

Bagley, M., & Leggett, J. (2021, December 6). *Marie Curie: Facts and biography.* Live Science. https://www.livescience.com/38907-marie-curie-facts-biography.html

*BBC - History - Marie Curie.* (2011, July 18). https://www.bbc.co.uk/history/historic_figures/curie_marie.shtml

Beheler, T. (2022, February 16). *Famous Couples: Marie and Pierre Curie | Headlines & Heroes.* The Library of Congress. https://blogs.loc.gov/headlinesandheroes/2022/02/famous-couples-marie-and-pierre-curie/

Bensaude-Vincent, B. (2026, February 12). *Dmitri Mendeleev*. Encyclopedia Britannica. https://www.britannica.com/biography/Dmitri-Mendeleev

Biography.com Editors. (2026, March 19). *Marie Curie.* Biography. https://www.biography.com/scientists/marie-curie

Britannica Editors (2022, December 6). *Curie.* Encyclopedia Britannica. https://www.britannica.com/science/curie

*Client challenge.* (n.d.). https://www.scribd.com/document/411873793/Adloff-The-Laboratory-Notebooks-of-Pierre-and-Marie-Curie

Coppack, L. (2024, January 8). *Bringing X-rays to the front.* British Society for the History of Medicine. https://bshm.org.uk/bringing-x-rays-to-the-front/?utm_source=copilot.com

*CURIE - The comic strip.* (n.d.). The Curie Method. Retrieved June 25, 2026, from https://lamethodecurie.fr/en/rubrique1.html?_gl=1*mjimdo*_up*MQ..*_ga*Mzc1MDc5MzMxLjE3NzMxNjg4NDE.*_ga_G7RMZZ61ZW*czE3NzMxNjg4NDAkbzEkZzEkdDE3NzMxNjg4NTMkajQ3JGwwJGgw*_ga_BRXE2LPE05*czE3NzMxNjg4NDAkbzEkZzEkdDE3NzMxNjg4NTMkajQ3JGwwJGgw*_ga_3H1QKXKJYY*czE3NzMxNjg4NDEkbzEkZzEkdDE3NzMxNjg4NTMkajQ4JGwwJGgw#19

*Curie – Unit of Radioactivity.* (2026). Nuclear Power. https://www.nuclear-power.com/nuclear-engineering/radiation-protection/units-of-radioactivity/curie-unit-of-radioactivity/

Curie, È. (1938). *Madame Curie.* Doubleday, Doran & Company.

Curie, P., M., Curie, M., & Bémont, G. (1898). *Sur une nouvelle substance fortement radio-active contenue dans la pechblende. C.R.*, *127*, 1215–1217. https://www.academie-sciences.fr/pdf/dossiers/Curie/Curie_pdf/CR1898_p1215_1217.pdf

*Curium - Element information, properties and uses | Periodic Table.* (n.d.). Retrieved June 24, 2026, from https://periodic-table.rsc.org/element/96/curium?utm_source=copilot.com

*December 1898: The Curies Discover Radium.* (2004, December 1). APS.org. https://www.aps.org/apsnews/2004/12/curies-1898-discover-radium

Developer, & Developer. (2025, March 28). *How are X-rays Produced?* Health Images. https://www.healthimages.com/how-x-rays-work/

Editorial Staff & Editorial Staff. (2021, May 10). *38 Marie Curie Facts: Interesting facts about Marie Curie.* The Fact File. https://thefactfile.org/marie-curie-facts/

English, T. (2019, September 28). *The amazing life and inventions of Marie Curie.* Interesting Engineering. https://interestingengineering.com/innovation/the-amazing-life-and-inventions-of-marie-curie

Goldsmith, B. (2005). *Obsessive Genius: The inner world of Marie Curie.* W. W. Norton & Company.

*Healing Energy: Radium in America.* (n.d.). Historical Medical Library. Retrieved June 24, 2026, from https://cppdigitallibrary.org/exhibits/show/radium/curieamerica?utm_source=copilot.com

History Sphere. (2025, October 4). *Pierre Curie, the quiet genius whose discoveries lit the path to modern physics and chemistry.* History Sphere – Histoire Du Monde Et Des Civilisations. https://historysphere.com/pierre-curie-1859/

History.com Staff. (2025, August 11). *Marie Curie: Facts about the Pioneering Chemist | HISTORY.* HISTORY. https://www.history.com/articles/marie-curie-facts

*How did the Curies Measure Radioactivity? - Google Arts & Culture.* (n.d.). Google Arts & Culture. Retrieved June 4, 2026, from https://artsandculture.google.com/story/how-did-the-curies-measure-radioactivity-mus%C3%A9e-curie/BwWx_qy57jACKQ?hl=en

*How the Curie Came to Be.* (n.d.). Wayback Machine. Retrieved June 4, 2026, from https://web.archive.org/web/20120320124750/http:/www.orau.org/ptp/articlesstories/thecurie.htm

*Image 7 of New-York Tribune (New York [N.Y.]), April 20, 1906.* (1906, April 20). Library of Congress. https://www.loc.gov/resource/sn83030214/1906-04-20/ed-1/?sp=7&q=Curie+CURIE+Curies+Pierre&r=-0.659,0,2.317,1.393,0

Jade, & Jade. (2026, February 23). *Marie Curie's affair scandal and her second Nobel.* History Things. https://historythings.com/marie-curie-affair-second-nobel-scandal/?utm_source=copilot.com

*Joachimsthal Uranium Mine.* (n.d.). Retrieved June 24, 2026, from http://large.stanford.edu/courses/2017/ph241/hasson2/

Johnson Lewis, J. (2019, March 16). *Marie Curie: mother of modern physics, researcher of radioactivity.* ThoughtCo. https://www.thoughtco.com/marie-curie-biography-3529555

*Józef Boguski.* (2025, January 1). Wiki 2. https://wiki2.org/en/J%C3%B3zef_Boguski

*Józef Boguski.* (n.d.). Academic Dictionaries and Encyclopedias. Retrieved June 24, 2026, from https://en-academic.com/dic.nsf/enwiki/509769

King, H. M. (n.d.). *Uraninite: A radioactive mineral and ore of uranium.* Retrieved June 24, 2026, from https://geology.com/minerals/uraninite.shtml

Kowalczyk, J. (2020, September 14). *Maria Skłodowska-Curie: The Nobel Pioneer.* CulturePL. https://culture.pl/en/article/maria-sklodowska-curie-the-nobel-pioneer

Krulwich, R. (2010, December 14). Don't come to Stockholm! Madame Curie's Nobel scandal. *NPR.*

https://www.npr.org/sections/krulwich/2010/12/14/132031977/don-t-come-to-stockholm-madame-curie-s-nobel-scandal

*Lost_Hospitals_of_London*. (n.d.). Retrieved June 24, 2026, from https://www.ezitis.myzen.co.uk/mariecurie.html

*Mariage d'Hélène Joliot et Michel Langevin, entourés d'Irène et Frédéric Joliot-Curie, novembre 1948 · Mariage d'Hélène Joliot et Michel Langevin, entourés d'Irène et Frédéric Joliot-Curie, novembre 1948 · Musée Curie · Musée Curie*. (n.d.). https://curiotheque.musee.curie.fr/s/en/item/28241?utm_source=copilot.com

*Marie and Pierre Curie and the discovery of polonium and radium*. (n.d.). NobelPrize.org. Retrieved June 24, 2026, from https://www.nobelprize.org/prizes/themes/marie-and-pierre-curie-and-the-discovery-of-polonium-and-radium/

*Marie Curie. Institut Curie*. (n.d.). Retrieved June 24, 2026, from https://mariecurie-usa.org/marie-curie/

*Marie Curie and the science of radioactivity*. (n.d.). Retrieved June 4, 2026, from https://history.aip.org/exhibits/curie/scandal1.htm

*Marie Curie and the science of radioactivity*. (n.d.-b). Retrieved June 4, 2026, from https://history.aip.org/exhibits/curie/recdis1.htm

*Marie Curie Biography - life, family, children, wife, school, mother, young, book, information, born, husband*. (n.d.). Retrieved June 24, 2026, from https://www.notablebiographies.com/Co-Da/Curie-Marie.html

*Marie Curie the scientist | Bio, facts & quotes*. (n.d.). Marie Curie. Retrieved June 24, 2026, from https://www.mariecurie.org.uk/about-us/our-history/marie-curie-the-scientist

*Marie Curie: a guide to the scientist's life, achievements and legacy*. (2025, July 1). HistoryExtra. https://www.historyextra.com/period/first-world-war/life-of-the-week-marie-curie/

*Mme. Curie is dead; martyr to science.* (n.d.). The New York Times. Retrieved June 4, 2026, from https://archive.nytimes.com/www.nytimes.com/learning/general/onthisday/bday/1107.html?pagewanted=all

*Nobel Prize in Physics 1903.* (n.d.). NobelPrize.org. Retrieved June 4, 2026, from https://www.nobelprize.org/prizes/physics/1903/marie-curie/biographical/

*Nobelist Marie Curie Has Affair with Physicist Paul Langevin | History | Research Starters | EBSCO Research.* (n.d.). EBSCO. Retrieved June 24, 2026, from https://www.ebsco.com/research-starters/history/nobelist-marie-curie-has-affair-physicist-paul-langevin

*Painlevé, Paul.* (n.d.). Book of Proofs. Retrieved June 24, 2026, from https://bookofproofs.github.io/history/19th-century/painleve.html?utm_source=copilot.com

Pastonesi, M. P. (2025, February 16). *Mealtime. Pierre, Marie And That Love Celebrated On Bicycle.* Tuttobiciweb. https://www.tuttobiciweb.it/article/2025/02/16/1739640218/cycling-history-science-honeymoon?lang=en

*Paul Appell.* (n.d.). Maths History. Retrieved June 24, 2026, from https://mathshistory.st-andrews.ac.uk/Biographies/Appell/

*Paul Painlevé.* (n.d.). Maths History. Retrieved June 24, 2026, from https://mathshistory.st-andrews.ac.uk/Biographies/Painleve/?utm_source=copilot.com

*Pierre and Marie Curie's honeymoon bicycle trip - Musée Curie (Coll. ACJC). - Google Arts & Culture.* (n.d.). Google Arts & Culture. Retrieved June 24, 2026, from https://artsandculture.google.com/asset/pierre-and-marie-curie%E2%80%99s-honeymoon-bicycle-trip-mus%C3%A9e-curie-coll-acjc/mQGutyig9hpjAg?hl=en

*Pierre Curie.* (2021, May 19). Biography. https://www.biography.com/scientists/pierre-curie

*Pierre Curie*. (2026, May 11). Britanica. https://www.britannica.com/biography/Pierre-Curie

Quinn, S. (2011). *Marie Curie: A life*. Plunkett Lake Press.

Redniss, L. (2015). *Radioactive*. Dey Street Books.

*Research Guides: French Women & Feminists in History: A resource Guide: Marie Curie*. (n.d.). Retrieved June 4, 2026, from https://guides.loc.gov/feminism-french-women-history/famous/marie-curie

Science History Institute. (2023, May 31). *Marie Curie, Marie Meloney, and The Significance of a gram of Radium | Science History Institute.* https://www.sciencehistory.org/stories/magazine/marie-curie-marie-meloney-and-the-significance-of-a-gram-of-radium/

Science History Institute. (2026, February 23). *Marie Sklodowska Curie | Science History Institute.* https://www.sciencehistory.org/education/scientific-biographies/marie-sklodowska-curie/

The Editors of Encyclopedia Britannica (2026, March 9). *Marie Curie summary. Encyclopedia Britannica.* https://www.britannica.com/summary/Marie-Curie

*The quadrant electrometer*. (n.d.). The Curie Method. Retrieved June 4, 2026, from https://lamethodecurie.fr/en/article23.html?_gl=1*1aads6q*_up*MQ..*_ga*MTMwNTMzOTg2Ny4xNzczNDEwODkz*_ga_G7RMZZ61ZW*czE3NzM0MTA4OTMkbzEkZzAkdDE3NzM0MTA4OTMkajYwJGwwJGgw*_ga_3H1QKXKJYY*czE3NzM0MTA4OTMkbzEkZzAkdDE3NzM0MTA4OTMkajYwJGwwJGgw*_ga_BRXE2LPE05*czE3NzM0MTA4OTMkbzEkZzAkdDE3NzM0MTA4OTMkajYwJGwwJGgw

The Tombstone Tourist. (2021, May 20). *Pierre Curie - The Tombstone Tourist.* https://thetombstonetourist.com/graves/pierre-curie/

Thompson, G. (n.d.). *Marie Curie: a pioneer in science and history.* worldhistory.org.uk. Retrieved June 24, 2026, from https://www.worldhistory.org.uk/scientists-and-inventors-marie-curie

Wikipedia contributors. (2025, October 10). *Kazimierz Żorawski.* Wikipedia. https://en.wikipedia.org/wiki/Kazimierz_%C5%BBorawski

Wojtowicz, T. (2024, August 20). *The All-Important question: What actually is a chronometer? - SwissWatches Magazine.* Swisswatches Magazine. Retrieved June 4, 2026, from https://swisswatches-magazine.com/what-is-a-chronometer/

WPI, Environmental Information Services -- Shawn Denny, Information Architect; Mike Pizzuti, Graphic Designer; Chelene Neal, Web Information Specialist; Kate Bessiere, Web Information Specialist. (n.d.). *What is radioactivity?* Retrieved June 24, 2026, from https://ehss.energy.gov/ohre/roadmap/achre/intro_9_2.html

Zz_Hugo. (2025, November 5). *Uranic rays.* radioactivity.eu.com. https://radioactivity.eu.com/articles/phenomenon/uranic_rays

## Chapter 2: Jane Goodall

"Jane Goodall." famousscientists.org. (2015, December 14). Web. https://www.famousscientists.org/jane-goodall/

Academy of Achievement. (2025, October 2). *Dame Jane Goodall | Academy of Achievement.* https://achievement.org/achiever/jane-goodall/

*Africa: For the first time.* (n.d.). Jane Goodall. Retrieved June 4, 2026, from https://janegoodallblog.weebly.com/africa-for-the-first-time.html

*Big Search On For Kidnaped Students.* (1975, May 22). California Digital Newspaper Collection. UCR. https://cdnc.ucr.edu/?a=d&d=DS19750522.2.6&e=-------en--20--1--txt-txIN--------

Bowe, J., & Bowe, J. (2025, October 31). *The scientific legacy of Dr. Jane Goodall.* The Earth Commons. https://earthcommons.georgetown.edu/all/the-scientific-legacy-of-dr-jane-goodall/

Bucher, A. (2025, October 1). *Jane Goodall was married twice: her husbands were Hugo van Lawick & Derek Bryceson.* Men's Journal. https://www.mensjournal.com/news/jane-goodall-was-married-twice-her-husbands-were-hugo-van-lawick-derek-bryceson

Burke, M. (2025, July 14). *"They hold hands, they embrace, they kiss": The woman who changed our view of chimps - and human beings.* https://www.bbc.com/culture/article/20250710-the-woman-who-redefined-chimps-and-human-beings

Butler, R. A. (2026, April 3). *Today is Jane Goodall Day. Her movement continues.* Mongabay. https://news.mongabay.com/2026/04/today-is-jane-goodall-day-her-movement-continues/

Caruso, C. (2025, October 1). Jane Goodall didn't have a college degree. she still changed everything we know about chimpanzees. *Biography.* https://www.biography.com/scientists/a68157702/jane-goodall-chimpanzees-expert

Caruso, C. (2025a, October 1). Jane Goodall. *Biography.* https://www.biography.com/scientists/jane-goodall

*Congo, Democratic Republic of (Zaire) | Encyclopedia.com.* (n.d.). Retrieved June 24, 2026, from https://www.encyclopedia.com/international/legal-and-political-magazines/congo-democratic-republic-zaire

Dan, & Dan. (2023, April 19). 40 Interesting Facts about Jane Goodall - World's Facts. *World's Facts - World's Interesting Facts.* https://www.worldsfacts.com/40-interesting-facts-about-jane-goodall/

*Early life.* (n.d.). Jane Goodall. Retrieved June 24, 2026, from https://janegoodallandthechimps.weebly.com/early-life.html

Famous Scientists. (2018, August 15). *Jane Goodall - biography, facts and pictures.* https://www.famousscientists.org/jane-goodall/

Friel, J. (2025, October 1). *Prosopagnosia (Face Blindness): Symptoms, Diagnosis & Famous People Who Have It.* Talk Nerdy to Me®. https://www.talknerdytome.com/blog/dsmrn5be2kkhvnjy5kcft29qi6x32r

Gacute, A. (2025, October 2). Dr. Jane Goodall's Family: Everything to Know After Her Death at 91. *OK Magazine.* https://okmagazine.com/p/jane-goodall-family-husband-sibling-son/

Golden, H. (2025, November 12). Inspirational chimpanzee expert Jane Goodall is laid to rest. *The Independent.* https://www.independent.co.uk/bulletin/news/jane-goodall-chimpanzee-expert-funeral-b2863840.html

Goodall, J. (1986). *The Chimpanzees of Gombe: Patterns of Behavior.* Belknap Press.

Goodall, J. (n.d.). *Extended bio.* Retrieved June 24, 2026 from https://www.janegoodall.org/wp-content/uploads/2020_Bio_Long_JaneGoodall.pdf

Greenfieldboyce, N. (2025, October 1). Jane Goodall, legendary primatologist, has died at age 91. *KPBS Public Media.* https://www.kpbs.org/news/science-technology/2025/10/01/jane-goodall-legendary-primatologist-has-died-at-age-91

*Honoring Jane Goodall's legacy of compassion for all animals.* (2026, April 3). Mercy for Animals. https://mercyforanimals.org/blog/jane-goodall-day/

Hooper, R. (2025, October 2). *How Jane Goodall changed the way we see animals – and the world.* New Scientist. https://www.newscientist.com/article/2498651-how-jane-goodall-changed-the-way-we-see-animals-and-the-world/

*Jane Goodall Biography - life, family, childhood, children, parents, wife, school, mother.* (n.d.). Encyclopedia World Biography. Retrieved June 24,

2026, from https://www.notablebiographies.com/Gi-He/Goodall-Jane.html

*Jane Goodall.* (2026, May 16). Britannica. https://www.britannica.com/biography/Jane-Goodall

*Jane Goodall.* (n.d.). National Geographic. Retrieved June 24, 2026, from https://education.nationalgeographic.org/resource/jane-goodall/

*Jane Goodall: The longest-running field study of primates.* (n.d.). Guinness World Records. Retrieved June 24, 2026, from https://www.guinnessworldrecords.com/news/icons/jane-goodall-longest-running-field-study-of-primates

*Judith Goodall (Judy): the sister of Jane Goodall & keeper of their childhood legacy - Celebz Wurld.* (2025, October 7). Celebz Wurld. https://celebzwurld.com/judith-goodall-judy-the-sister-of-jane-goodall-keeper-of-their-childhood-legacy/

Korrai, Γ. E. (n.d.). *Jane Goodall Biography: How one woman changed the way we see animals.* Retrieved on June 24, 2026 from https://maxmag.org/jane-goodall-biography-life-with-the-chimpanzees/

Krauser, E. (2025, October 1). *All about Jane Goodall's son, Hugo "Grub" Eric Louis Van Lawick.* People. https://people.com/all-about-jane-goodall-son-8727343

*Louis Leakey, Anthropologist born - African American Registry.* (2025, August 7). African American Registry. https://aaregistry.org/story/louis-leakey-anthropologist-born/

Magazine, B. W. (2026, June 15). *Who was Jane Goodall? We explore the life of the groundbreaking conservationist and how she changed the world.* Discover Wildlife. https://www.discoverwildlife.com/people/jane-goodall-facts

*Mary Leakey Biography - life, family, childhood, children, parents, death, school, mother.* (n.d.). https://www.notablebiographies.com/Ki-Lo/Leakey-Mary.html

*Mortimer Morris-Goodall: The racing Father & Early Influence on Jane Goodall - Celebz Wurld.* (2025, October 4). Celebz Wurld. https://celebzwurld.com/mortimer-morris-goodall-the-racing-father-early-influence-on-jane-goodall/

MSN. (n.d.). *April 3, 1934 - Primatologist Jane Goodall born in London.* Retrieved June 24, 2026, from https://www.msn.com/en-us/science/general/april-3-1934-primatologist-jane-goodall-born-in-london/ar-AA2057AY

MSN. *Jane Goodall: A life with the wild* (n.d.). Retrieved June 24, 2026, from https://www.msn.com/en-us/lifestyle/lifestyle-buzz/jane-goodall-a-life-with-the-wild/ss-AA1pIh3B#image=1

*My Life with the Chimpanzees Study Guide.* (n.d.). SuperSummary. Retrieved June 4, 2026 from https://www.supersummary.com/my-life-with-the-chimpanzees/summary/

National Park. (2026, May 9). *Gombe Stream National Park | (Official GANP Page).* (Official GANP Page). https://national-parks.org/tanzania/gombe-stream/

Online, E. (2025, October 2). Was Jane Goodall married and how rich was late chimpanzee expert? Exploring her family, fortune and cause. *The Economic Times.* https://economictimes.indiatimes.com/news/international/global-trends/us-news-was-jane-goodall-married-and-how-rich-was-late-chimpanzee-expert-exploring-her-family-fortune-and-cause-of-death/articleshow/124274232.cms

Osborne, M. (2025, October 1). Jane Goodall, legendary primatologist and anthropologist, dies at 91. *Smithsonian Magazine.* https://www.smithsonianmag.com/science-nature/jane-goodall-legendary-primatologist-and-anthropologist-dies-at-91-180984251/

Parul, P. (2025, October 1). Jane Goodall Death: Who were her two husbands, kids, career | and why she never remarried. *The Daily Guardian.* https://thedailyguardian.com/top-news/jane-goodall-death-who-were-her-two-husbands-kids-career-and-why-she-never-remarried-667246/

Rai, S. (2026, March 9). *Jane Goodall's Family: All About Her Husband and Kids*. GazetteDirect. https://gazettedirect.com/jane-goodall-husband-and-son/

*Roots & Shoots: a global movement of empowered youth.* (2025). Jane Goodall Institute. https://janegoodall.ca/wp-content/uploads/2025/06/2025-Fact-Sheet-Roots-Shoots-Global.pdf?utm_source=copilot.com

Staff, T. (2026, April 3). Jane Goodall interview in Famous Last Words Netflix Documentary. *Netflix Tudum.* https://www.netflix.com/tudum/articles/jane-goodall-famous-last-words-documentary

Sundel, J., & Whisnant, G. (2025, October 20). Jane Goodall's official cause of death revealed. *Newsweek.* https://www.newsweek.com/jane-goodall-official-cause-death-cardiac-arrest-10907773

Temin, D. (2026, April 4). *What Jane Goodall is teaching us about leaving a legacy.* Forbes. https://www.forbes.com/sites/daviatemin/2026/04/03/what-jane-goodall-is-teaching-us-about-leaving-a-legacy/

*The evolution of Dr. Jane Goodall.* (n.d.). Natural History Museum. Retrieved June 24, 2026, from https://nhm.org/stories/evolution-dr-jane-goodall

*The Jane Goodall Institute Homepage.* (2026, May 12). Jane Goodall Institute USA. https://janegoodall.org/

*The Jane Goodall Institute New Zealand.* (n.d.). Retrieved June 24, 2026, from https://www.janegoodall.org.nz/africa-programmes/all-about-chimps/chimpanzee-communication-1/?utm_source=copilot.com

The Leakey Foundation. (2025, April 30). *Episode 03: Jane Goodall - The Leakey Foundation.* https://leakeyfoundation.org/podcast/episode-03-jane-goodall/

*The Westport Library Resource Guides: Jane Goodall: about.* (n.d.). Retrieved June 24, 2026 from https://westportlibrary.libguides.com/JaneGoodall

Tilden, A. (2026, April 3). *A reminder that ordinary people can change the world: A tribute to Jane Goodall — Keep Pinellas beautiful.* Keep Pinellas Beautiful. https://www.kpbcares.org/kpb-blog/2026/4/1/how-jane-goodall-reminded-all-of-us-that-ordinary-people-can-change-the-world

Torregiani, I. (2025, October 1). *Jane Goodall's Only Son Grew Up in the Wild — But He Found a Different Calling.* Parade. https://parade.com/news/jane-goodalhttps://parade.com/news/jane-goodalls-only-son-grew-up-in-the-wild-but-he-found-a-different-callingls-only-son-grew-up-in-the-wild-but-he-found-a-different-calling

Trivedi, N. (2025, October 1). Meet David Greybeard: Dr. Jane Goodall's favorite chimpanzee who changed animal science forever. *TimesNow.* https://www.timesnownews.com/world/us/us-buzz/meet-david-greybeard-dr-jane-goodalls-favorite-chimpanzee-who-changed-animal-science-forever-article-152928143

Vance, C. (2026, April 7). April 3 marks first annual Jane Goodall Day. *April 3 Marks First Annual Jane Goodall Day - the Post.* https://www.thepostathens.com/article/2026/04/jane-goodall-day-caitlyn-vance

*Vanne Morris Goodall: The Mother and Quiet Heart Behind Jane Goodall's Ambition & Legacy (Margaret Myfanwe Joseph) - Celebz Wurld.* (2025, October 3). Celebz Wurld. https://celebzwurld.com/vanne-morris-goodall-the-quiet-heart-behind-jane-goodalls-ambition-legacy-margaret-myfanwe-joseph/

Werner, A. (2022, August 2). *10 famous female scientists who wrote history.* Integra. https://www.integra-biosciences.com/united-states/en/blog/article/famous-female-scientists?msclkid=42a2a7e1a0b91a56ee2ace37047fbbe5&utm_campaign=Blog%20US%20%E2%80%93%20EN&utm_content=Blog%C2%A09%3A%20Antimicrobial%20Susceptibility%20Testing&utm_medium=cpc&utm_source=bing&utm_term=Jane%20Goodall

Wikipedia contributors. (2026, June 20). *Jane Goodall.* Wikipedia. https://en.wikipedia.org/wiki/Jane_Goodall

Wikipedia contributors. (2026, March 1). *Gombe National Park.* Wikipedia. https://en.wikipedia.org/wiki/Gombe_National_Park

Williets, B. (2022, July 12). Dr Jane Goodall: the life work of the local conservationist. *Bournemouth Echo.* https://www.bournemouthecho.co.uk/news/19309250.dr-jane-goodall-life-work-local-conservationist/

## Chapter 3: Ada Lovelace

Abby. (2025, May 23). *Ada Lovelace – complete biography, history and inventions.* History-Computer. https://history-computer.com/people/ada-lovelace-complete-biography/

*About Ada Lovelace.* (2015, September 10). Ada Lovelace. https://blogs.bodleian.ox.ac.uk/adalovelace/about-ada-lovelace/

Academy, T. (2025, November 18). Ada Lovelace: the brilliant mind who saw the future of computers. *Think Academy U.S. Blog -.* https://www.thethinkacademy.com/blog/edubriefs-ada-lovelace-the-brilliant-mind-who-saw-the-future-of-computers/

*Ada Lovelace.* Computing history. (n.d.). Retrieved June 24, 2026 from https://www.computinghistory.org.uk/det/1843/Ada-Lovelace/

*Ada Lovelace.* (2023, November 28). Historic UK. https://www.historic-uk.com/CultureUK/Ada-Lovelace/

*Ada Lovelace | Lemelson.* (n.d.). Retrieved on June 14, 2026, from https://lemelson.mit.edu/resources/ada-lovelace

*Ada Lovelace and the first computer programme in the world.* (n.d.). Max-Planck-Gesellschaft. Retrieved on June 24, 2026 from https://www.mpg.de/female-pioneers-of-science/Ada-Lovelace

*Ada Lovelace blazed a trail in science – we need more women to follow in her footsteps*. (2016, October 11). The Conversation. https://theconversation.com/ada-lovelace-blazed-a-trail-in-science-we-need-more-women-to-follow-in-her-footsteps-66661

*Ada Lovelace changed computing one hundred years after her death*. (n.d.). Sky HISTORY TV Channel. Retrieved June 24, 2026 from https://www.history.co.uk/articles/ada-lovelace-changed-computing-one-hundred-years-after-her-death

Ada Lovelace Institute. (n.d.). Retrieved June 24, 2026 from https://www.adalovelaceinstitute.org/

*Ada Lovelace Project - Rhineland-Palatinate Competence Center for Women in STEM*. (n.d.). Landeshauptstadt Mainz. Retrieved on June 24, 2026 from https://www.mainz.de/en/verzeichnisse/vereinsverzeichnis/Ada-Lovelace-Projekt_-_Mentoring-Netzwerk_fuer_Maedchen_und_junge_Frauen_in_MINT_

*Ada Lovelace*. (2018, October 9). ETHW. https://ethw.org/Ada_Lovelace

*Ada Lovelace*. (n.d.). Britannica. Retrieved June 24, 2026, from https://www.britannica.com/biography/Ada-Lovelace

*Ada Lovelace*. (n.d.). Maths History. Retrieved June 24, 2026 from https://mathshistory.st-andrews.ac.uk/Biographies/Lovelace/

*Ada Lovelace*. (n.d.). Retrieved June 24, 2026 from https://scientificwomen.net/women/lovelace-ada-59

*Ada Lovelace: a visionary of computing*. (2024, October 20). HistoryExtra. https://www.historyextra.com/period/modern/ada-lovelace-stem-women-computing-science-facts-life/

*Ada Lovelace's Mathematical Papers - Clay Mathematics Institute*. (2025, February 12). Clay Mathematics Institute. https://www.claymath.org/online-resources/ada-lovelaces-mathematical-papers/

Aiello, L. C. (2016). The multifaceted impact of Ada Lovelace in the digital age. *Artificial Intelligence*, *235*, 58–62. https://www.sciencedirect.com/science/article/pii/S0004370216300224

*BBC Radio 4 Extra - The Letters of Ada Lovelace, The Poetry of Mathematics.* (2015, September 14). BBC. https://www.bbc.co.uk/programmes/b069jjmg

BBC. (2015, September 15). *Calculating Ada: The Countess of Computing - Not your typical role model: Ada Lovelace the 19th century programmer.* https://www.bbc.co.uk/programmes/articles/3jNQLTMrPlYGTBn0WV6M2MS/not-your-typical-role-model-ada-lovelace-the-19th-century-programmer

Bevan, N. (2026, March 17). *Secret East Horsley tunnel to reopen after being hidden for 50 years.* https://www.bbc.com/news/articles/cpd8w1dpg0wo

Brown, M. (2009, March 24). *The Scientific Tourist in London: #3 Ada Lovelace Plaque | nature.com blogs.* https://blogs.nature.com/blog/the-scientific-tourist-in-london-3-ada-lovelace-plaque/

Demming, A. (n.d.). *Ada Lovelace.* New Scientist. Retrieved June 24, 2026, from https://www.newscientist.com/people/ada-lovelace/

Dobbs, G. (2014, September 25). *Behind the lit: Lord Byron's visionary, Opiate-Addicted, Hard-Gambling daughter — the Airship.* The Airship. http://airshipdaily.com/blog/06112014-ada-lovelace

*Gambling in Regency and Victorian England.* (n.d.). The Victorian Web. Retrieved on June 4, 2026, from https://victorianweb.org/history/leisure/gambling1.html

Garvin, K. (n.d.). *Chapter 12 – Ada Lovelace.* History of Applied Science & Technology. Retrieved June 24, 2026 from https://press.rebus.community/historyoftech/chapter/ada-lovelace/

Government Art Collection. (2026, March 18). *The legacy of Ada Lovelace - Government Art Collection.* https://artcollection.dcms.gov.uk/stories/the-legacy-of-ada-lovelace/

History, H. (2021, July 9). *Ada Lovelace.* Hourly History. https://hourlyhistory.com/ada-lovelace/

Hollings, C., Martin, U., & Rice, A. (2017). The early mathematical education of Ada Lovelace. *BSHM Bulletin Journal of the British Society for the History of Mathematics, 32*(3), 221–234. https://doi.org/10.1080/17498430.2017.1325297

Johnson Lewis, J. (2019b, October 17). *Biography of Ada Lovelace.* ThoughtCo. https://www.thoughtco.com/ada-lovelace-biography-3525491

Johnson, L. (n.d.). *10 Facts about Ada Lovelace: the first computer Programmer.* History Hit. Retrieved June 24, 2026 from https://www.historyhit.com/facts-about-ada-lovelace-the-first-computer-programmer/

Klein, C. (2025, May 27). *10 things You may not know about Ada Lovelace | HISTORY.* HISTORY. https://www.history.com/articles/10-things-you-may-not-know-about-ada-lovelace

Longley, R. (2021, February 19). *Biography of Ada Lovelace, First Computer Programmer.* ThoughtCo. https://www.thoughtco.com/ada-lovelace-biography-5113321

MacFarlane, A. (2013, May 15). *Ada Lovelace (1815-1852).* Philosophy Now. https://philosophynow.org/issues/96/Ada_Lovelace_1815-1852

Martin, U. (2018, July 27). *Ada Lovelace: the Making of a Computer Scientist.* Bodleian Libraries. University of Oxford. https://blogs.bodleian.ox.ac.uk/adalovelace/

*Mathematical Treasure: Ada Lovelace's notes on the Analytic Engine | Mathematical Association of America.* (n.d.). Retrieved June 4, 2026

from https://old.maa.org/press/periodicals/convergence/mathematical-treasure-ada-lovelaces-notes-on-the-analytic-engine

*Medora Leigh.* (n.d.). Lord Byron and his Times. Retrieved on June 4, 2026, from https://www.lordbyron.org/persRec.php?&selectPerson=MeLeigh1849

MILLER, C. C. (2026). *Ada Lovelace.* The New York Times. https://www.nytimes.com/interactive/2018/obituaries/overlooked-ada-lovelace.html

Morais, B. (2013, October 15). Ada Lovelace, the first tech visionary. *The New Yorker.* https://www.newyorker.com/tech/annals-of-technology/ada-lovelace-the-first-tech-visionary

News. (2024, October 15). *Lovelace: The programmer who spooked Alan Turing.* Mind Matters. https://mindmatters.ai/2020/05/lovelace-the-programmer-who-spooked-alan-turing/

Padua, S. (n.d.). *Who was Ada?* Ada Lovelace Day. Retrieved June 24, 2026, from https://findingada.com/about/who-was-ada/

Pridmore, O. (2024, August 23). Lord Byron's resting place to be "safeguarded for future" under major new plans. *Nottingham Post.* https://www.nottinghampost.com/news/local-news/lord-byrons-resting-place-safeguarded-9502620

Stannard, R. (2026, March 19). *What did the term "Computer" mean to Ada Lovelace?* AEANET. https://www.aeanet.org/what-did-the-term-computer-mean-to-ada-lovelace/

Stein, D. (1987). *Ada: A Life and a Legacy.* Mit Press.

Taylor, A. (2020). Ada Lovelace, pioneer. *JSTOR Daily.* https://daily.jstor.org/ada-lovelace-pioneer/

The History Press. (2024, April 25). *Charles Babbage and Ada Lovelace: The computer's most passionate partnership - The History Press.* https://thehistorypress.co.uk/article/charles-babbage-and-ada-lovelace-the-computers-most-passionate-partnership/

*The life of Ada Lovelace.* (2019, February 9). History Today. https://www.historytoday.com/archive/months-past/life-ada-lovelace

*The mathematics of Ada Lovelace | Mathematical Institute.* (n.d.). Retrieved from Jun 24, 2026 from https://www.maths.ox.ac.uk/groups/history-mathematics/mathematics-ada-lovelace

Toole, B. (n.d.). *Ada Byron, Lady Lovelace (1815-1852).* cs.yale.edu. Retrieved June 24, 2026, from https://cs.yale.edu/homes/tap/Files/ada-bio.html

Toole, B. A. (2010). *Ada, the enchantress of numbers: Poetical Science.* Betty Alexandra Toole.

*University of Southampton Ada Lovelace Excellence Scholarships.* (2026, March 16). Study in UK. https://www.studyinuk.com/university-of-southampton-ada-lovelace-excellence-scholarships/

Wolfram, S. (2015, December 10). *Untangling the tale of Ada Lovelace.* https://writings.stephenwolfram.com/2015/12/untangling-the-tale-of-ada-lovelace/

Zielinski, S. (2013, November 16). Who was Ada Lovelace? *Smithsonian Magazine.* https://www.smithsonianmag.com/science-nature/who-was-ada-lovelace-25830687/

Zwolak, J. (2025, February 11). *Ada Lovelace: the world's first computer programmer who predicted artificial intelligence.* NIST. https://www.nist.gov/blogs/taking-measure/ada-lovelace-worlds-first-computer-programmer-who-predicted-artificial

## Chapter 4: Rosalind Franklin

*A Science Odyssey: People and Discoveries: Rosalind Franklin.* (n.d.). Retrieved June 14, 2026, from https://www.pbs.org/wgbh/aso/databank/entries/bofran.html

Ashcroft, F. (2015). Rosalind Franklin and Photograph 51. *The Lancet, 386*(10002), 1439–1440.

https://www.thelancet.com/journals/lancet/article/PIIS0140-6736(15)00422-5/fulltext

Bagley, M. (2013, September 19). *Rosalind Franklin: Biography & Discovery of DNA Structure.* Live Science. https://www.livescience.com/39804-rosalind-franklin.html

Barbara Pfeffer Billauer JD MA (Occ. Health) PhD & Barbara Pfeffer Billauer JD MA (Occ. Health) PhD. (2025, April 11). Great Women of Science: Rosalind Franklin, hidden giant behind modeling of DNA. *American Council on Science and Health.* https://www.acsh.org/news/2025/04/10/great-women-science-rosalind-franklin-hidden-giant-behind-modeling-dna-49411

BBC News. (2020, July 25). *Rosalind Franklin centenary: "She would have been totally amazed."* https://www.bbc.com/news/uk-england-cambridgeshire-53514812

Benderly, B. L. (2018, August 1). *Rosalind Franklin and the damage of gender harassment.* Science. https://www.science.org/content/article/rosalind-franklin-and-damage-gender-harassment

Biographical overview. (n.d.). *Rosalind Franklin - Profiles in Science.* National Library of Medicine. Retrieved June 24, 2026, from https://profiles.nlm.nih.gov/spotlight/kr/feature/biographical

*Biography 19: Rosalind Elsie Franklin (1920-1958):* CSHL DNA Learning Center. (n.d.). https://dnalc.cshl.edu/view/16439-Biography-19-Rosalind-Elsie-Franklin-1920-1958-.html

BiologyInsights Team. (2025, July 26). *Dr. Rosalind Franklin: Her Contributions to DNA & Science - Biology Insights.* Biology Insights. https://biologyinsights.com/dr-rosalind-franklin-her-contributions-to-dna-science/

Burakoff, M. (2023, April 25). *Rosalind Franklin's role in DNA discovery gets a new twist.* AP News. https://apnews.com/article/dna-double-helix-rosalind-franklin-watson-crick-69ec8164c720e0b23374da69a1d3708d

Cobb, M. (2018, February 14). Sexism in science: did Watson and Crick really steal Rosalind Franklin's data? *The Guardian.* https://www.theguardian.com/science/2015/jun/23/sexism-in-science-did-watson-and-crick-really-steal-rosalind-franklins-data

Cobb, M., & Comfort, N. (2023). What Rosalind Franklin truly contributed to the discovery of DNA's structure. *Nature, 616*(7958), 657–660. https://doi.org/10.1038/d41586-023-01313-5

Conlon, A. M. (2020, August 3). *Rosalind Franklin.* New Scientist. https://www.newscientist.com/people/rosalind-franklin/

*Discovery of a lifetime.* (n.d.). Rosalind Franklin University. Retrieved June 24, 2026, from https://www.rosalindfranklin.edu/rf100/discovery.html

*DPMA | Rosalind Franklin.* (2026, March 3). Deutsches Patent- Und Markenamt. https://www.dpma.de/english/our_office/publications/ingeniouswomen/rosalindfranklin/index.html

Elkin, L. (n.d.). Rosalind Elsie Franklin. *Jewish Women's Archive.* Retrieved June 24, 2026, from https://jwa.org/encyclopedia/article/franklin-rosalind

lkin, L. O. (2003). Rosalind Franklin and the double Helix. *Physics Today, 56*(3), 42–48. https://doi.org/10.1063/1.1570771

Finnegan, M. (2021, December 3). *Meet Rosalind Franklin, a sidelined figure in the history of DNA science.* PBS. https://www.pbs.org/newshour/science/meet-rosalind-franklin-a-sidelined-figure-in-the-history-of-dna-science

Franklin, R. E. (1950). Influence of the bonding electrons on the scattering of X-Rays by carbon. *Nature, 165*(4185), 71–72. https://doi.org/10.1038/165071a0

*Franklin's X-ray diffraction, explanation of X-ray pattern.* :: CSHL DNA Learning Center. (n.d.). Retrieved June 24, 2026, from https://dnalc.cshl.edu/view/15014-Franklin-s-X-ray-diffraction-explanation-of-X-ray-pattern-.html

*From molecular structures to research culture: Celebrating Rosalind Franklin's legacy.* Royal Society. (2024, December 10). https://royalsociety.org/blog/2025/03/rosalind-franklin-award/

*From the archive: Rosalind Franklin's famous Photo 51.* (2023, April 25). UK Research and Innovation. https://www.ukri.org/blog/from-the-archive-rosalind-franklins-famous-photo-51/

Garman, E. F. (2020, July 3). *Rosalind Franklin 1920–1958.* IUCr Newsletter. https://www.iucr.org/news/newsletter/volume-28/number-2/rosalind-franklin-19201958

Gaudelli, N. M., & Komor, A. C. (2020). Celebrating Rosalind Franklin's Centennial with a Nobel Win for Doudna and Charpentier. *Molecular Therapy, 28*(12), 2519–2520. https://doi.org/10.1016/j.ymthe.2020.11.013

Glynn, J. (2012). Remembering my sister Rosalind Franklin. *The Lancet, 379*(9821), 1094–1095. https://www.thelancet.com/journals/lancet/article/PIIS0140-6736(12)60452-8/fulltext

How Rosalind Franklin was let down by DNA's dysfunctional team. (2023). *Nature, 616*(7958), 630. https://doi.org/10.1038/d41586-023-01390-6

Johnson, B. (2025, September 2). Rosalind Franklin's contributions to virology. *Research Communities by Springer Nature.* https://communities.springernature.com/posts/rosalind-franklin-s-contributions-to-virology

Kaiser, J., Meier, A., Collins, L., Parker, J., Singhal, A., & Sun, S. (n.d.). *Rosalind Elsie Franklin.* Breaking Barriers: Diversity and Equity in Chemistry. Retrieved June 24, 2026, from https://opentext.ku.edu/deichemistry/chapter/rosalind-elsie-franklin/?utm_source=copilot.com

Kennett, S. (n.d.). *Women in Science - Rosalind Franklin.* Science Café. Retrieved June 24, 2026, from https://sciencecafe.co.uk/post/rosalind-franklin

King's College London. (2020, July 20). *Rosalind Franklin.* https://www.kcl.ac.uk/people/rosalind-franklin

King's College London. (n.d.). *Rosalind Franklin: How her legacy lives on.* Retrieved June 4, 2026, from https://www.kcl.ac.uk/rosalind-franklin-100-anniversary-how-her-legacy-lives-on

Martins, K., & Fry, E. &. (2025). Rosalind Franklin: Equal Co-Discoverer of DNA? *World History Encyclopedia.* https://www.worldhistory.org/article/2845/rosalind-franklin/

Nick. (2022, May 19). *Rosalind Franklin: How she changed the world.* The History Ace. https://thehistoryace.com/rosalind-franklin-how-she-changed-the-world/

Oreskes, N. (2025, May 30). Rosalind Franklin deserves a posthumous Nobel Prize for co-discovering DNA structure. *Scientific American.* https://www.scientificamerican.com/article/rosalind-franklin-deserves-a-posthumous-nobel-prize-for-co-discovering-dna-structure/

Pederson, T. (2020). The double helix: "Photo 51" revisited. *The FASEB Journal, 34*(2), 1923–1927. https://doi.org/10.1096/fj.202000119

Percec, V. P., & Xiao, Q. (2021). The legacy of Rosalind E. Franklin: Landmark contributions to two Nobel Prizes. *ScienceDirect, 7*(3), 529–536. https://www.sciencedirect.com/science/article/pii/S2451929421001005

Phelan, L. (2021, July 8). Rosalind Franklin. *The Biophysical Society.* https://www.biophysics.org/profiles/rosalind-franklin

Pruitt, S. (2025, November 10). *Rosalind Franklin's role in the discovery of DNA's Structure.* HISTORY. https://www.history.com/articles/rosalind-franklin-dna-discovery

*Rosalind Franklin | Biography & Contributions to Science.* (n.d.). Study.com. Retrieved June 24, 2026, from https://study.com/buy/academy/lesson/scientist-rosalind-franklin-biography-quotes.html?src=ppc_bing_nonbrand&rcntxt=aws&crt=&kwd=SEO-PPC-BUY&kwid=dat-2329040505886940:loc-190&agid=1235851302596746&mt=b&device=c&network=o&_campaign=SeoPPC&msclkid=89fe3c0b4afc1a42183755dd1e931e92

*Rosalind Franklin | History | Research Starters | EBSCO Research.* (n.d.). EBSCO. Retrieved June 24, 2026, from https://www.ebsco.com/research-starters/history/rosalind-franklin

Rosalind Franklin Institute. (2025, March 12). *Rosalind Franklin's Life - Rosalind Franklin Institute.* https://www.rfi.ac.uk/discover-learn/rosalind-franklins-life/

*Rosalind Franklin University.* (2026, May 29). Rosalind Franklin University. https://www.rosalindfranklin.edu/

Rosalind Franklin was so much more than the 'wronged heroine' of DNA. (2020). *Nature, 583*(7817), 492. https://doi.org/10.1038/d41586-020-02144-4

Rosalind Franklin. (2020, June 15). *Biography.* https://www.biography.com/scientists/rosalind-franklin

*Rosalind Franklin.* (2026, May 22). Britannica. https://www.britannica.com/biography/Rosalind-Franklin

*Rosalind Franklin.* (n.d.). Jewish Virtual Library. Retrieved June 14, 2026, from https://jewishvirtuallibrary.org/rosalind-franklin

*Rosalind FRANKLIN*. (n.d.). History of Scientific Women. Retrieved June 24, 2026, from https://scientificwomen.net/women/franklin-rosalind-38

*Rosalind Franklin: A crucial contribution.* Learn Science at Scitable. (n.d.). Retrieved June 24, 2026, from https://www.nature.com/scitable/topicpage/rosalind-franklin-a-crucial-contribution-6538012/

*Rosalind Franklin: The Dark Lady of DNA*. (2023, June 14). National Air and Space Museum. https://airandspace.si.edu/stories/editorial/rosalind-franklin-dark-lady-dna

*Rosalind Franklin: the overlooked discoverer of DNA*. (n.d.). Rejected Princesses. Retrieved June 14, 2026, from https://www.rejectedprincesses.com/princesses/rosalind-franklin

*Rosalind Franklin: The unsung hero of the DNA double helix*. (2025, August 19). Bayer. https://www.bayer.com/en/rosalind-franklin-the-unsung-hero-of-the-dna-double-helix

Science History Institute. (2026b, February 24). *Francis Crick, Rosalind Franklin, James Watson, and Maurice Wilkins | Science History Institute*. https://www.sciencehistory.org/education/scientific-biographies/francis-crick-rosalind-franklin-james-watson-and-maurice-wilkins/

Science Insights Team. (2026, March 3). *Who was Rosalind Franklin and what did she discover?* ScienceInsights. https://scienceinsights.org/who-was-rosalind-franklin-and-what-did-she-discover/#google_vignette

Staff, S. (2020, October 13). Who is Rosalind Franklin? *ScienceAlert*. https://www.sciencealert.com/rosalind-franklin

Stasiak, A. (2001). Rosalind Franklin. *EMBO Reports*, *2*(3), 181. https://doi.org/10.1093/embo-reports/kve037

Sullivan, W. (2023, May 1). What We're Still Learning About Rosalind Franklin's Unheralded Brilliance. *Smithsonian Magazine.* https://www.smithsonianmag.com/smart-news/what-were-still-learning-about-rosalind-franklins-unheralded-brilliance-180982058/

*The Holes in Coal: Research at BCURA and in Paris, 1942-1951.* (n.d.). Rosalind Franklin - Profiles in Science. Retrieved June 4, 2026, from https://profiles.nlm.nih.gov/spotlight/kr/feature/coal

*The papers of Rosalind Franklin.* (n.d.). Retrieved June 24, 2026, from Wellcome Collection. https://wellcomecollection.org/works/ka25u4ft

*The remarkable life of Rosalind Franklin.* (n.d.). Sky History TV Channel. Retrieved June 4, 2026, from https://www.history.co.uk/articles/rosalind-franklin-the-dark-lady-of-dna

Thompson, B., & Bundell, S. (2023). How Rosalind Franklin's story was rewritten. *Nature.* https://doi.org/10.1038/d41586-023-01443-w

Werner, A. (2022b, August 2). *10 famous female scientists who wrote history.* Integra. https://www.integra-biosciences.com/united-states/en/blog/article/famous-female-scientists?msclkid=32e95f17aa181ee799f236c1fe77c118&utm_campaign=Blog%20US%20%E2%80%93%20EN&utm_content=Blog%C2%A09%3A%20Antimicrobial%20Susceptibility%20Testing&utm_medium=cpc&utm_source=bing&utm_term=rosalind%20franklin

Whitney, A. (2022, October 6). *15 Facts about Rosalind Franklin.* Mental Floss. https://www.mentalfloss.com/science/biology/15-overlooked-facts-about-rosalind-franklin

Wikipedia contributors. (2002, September 24). *Rosalind Franklin.* Wikipedia. https://en.wikipedia.org/wiki/Rosalind_Franklin

Wikipedia contributors. (n.d.). *File:Photo 51 x-ray diffraction image.jpg - Wikipedia.* Retrieved June 24, 2026, from https://en.wikipedia.org/w/index.php?curid=38068629

*Women in radiation History: Rosalind Franklin | U.S. EPA.* (2026, February 19). US EPA. https://www.epa.gov/radtown/women-radiation-history-rosalind-franklin

World History Edu. (2024, May 23). *Life and major accomplishments of Rosalind Franklin.* https://worldhistoryedu.com/life-and-major-accomplishments-of-rosalind-franklin/

## Chapter 5: Vera Rubin

*About | Rubin Observatory.* (n.d.). Retrieved June 14, 2026, from https://rubinobservatory.org/about

Bahcall, N. A. (2021). *Vera C Rubin.* National Academy of Sciences. https://www.nasonline.org/wp-content/uploads/2024/06/rubin-vera.pdf?utm_source=copilot.com

Boyle, A. (2026, April 3). *Rubin Observatory team discovers 11,000 new asteroids, with help from University of Washington software.* Geekwire. https://www.geekwire.com/2026/rubin-observatory-11000-new-asteroids/

Brian, A. (2026, June 20). *Vera Rubin.* BUHAVE. https://www.buhave.com/biography/vera-rubin/

Childers, T. (2019, June 11). *Vera Rubin: the astronomer who brought dark matter to light.* Space. https://www.space.com/vera-rubin.html

Cosmic Times. (n.d.). *Galaxies Still Misbehaving.* Retrieved June 24, 2026, from https://imagine.gsfc.nasa.gov/educators/programs/cosmictimes/educators/guide/1965/misbehaving.html

Gamillo, E., & Gamillo, E. (2023, December 26). *Vera Rubin found a lifetime of wonder in the dark skies.* Astronomy Magazine. https://www.astronomy.com/science/vera-rubin-found-a-lifetime-of-wonder-in-the-dark-skies/

Garofalo, D. (2024, March 7). On Vera Rubin: A life. *DavidGarofalosCorner*. https://www.davidgarofaloscorner.com/post/on-vera-rubin-a-life

Gisonna, K. (2026, March 18). *Vera Rubin*. Britannica. https://www.britannica.com/biography/Vera-Rubin

Larsen, K. (n.d.). Vera Cooper Rubin. *Jewish Women's Archive*. Retrieved June 24, 2026, from https://jwa.org/encyclopedia/article/rubin-vera-cooper

Mathewson, S. (2026, April 5). The powerful new Rubin Observatory just found 11,000 new asteroids and measured 'tens of thousands more' *Space*. https://www.space.com/astronomy/asteroids/the-powerful-new-rubin-observatory-just-found-11-000-new-asteroids-and-measured-tens-of-thousands-more

*NSF-DOE Vera C. Rubin Observatory*. (2025, June 23). NSF - U.S. National Science Foundation. https://www.nsf.gov/focus-areas/astronomy-space/rubin-observatory

*Phi Beta Kappa*. (2026, June 3). Britannica. https://www.britannica.com/topic/Phi-Beta-Kappa

*PPD The Women behind the science: Vera Rubin*. (n.d.). Retrieved June 4, 2026, from https://www.ppd.stfc.ac.uk/Pages/the-women-behind-the-science-vera-rubin.aspx

Rekik, K. (2026, April 4). Vera C. Rubin Observatory discovers 11,000 asteroids in just weeks, and this is only the start. *The Daily Galaxy - Great Discoveries Channel*. https://dailygalaxy.com/2026/04/vera-c-rubin-observatory-11000-asteroids/

*Rotating galaxies and dark matter*. (n.d.). Retrieved June 24, 2026, from https://pwg.gsfc.nasa.gov/stargaze/Sun4Adop3.htm

Rubin, Vera. C., Ford, W. K., Jr, & Thonnard, N. (1978). Extended Rotation Curves of High-Luminosity Spiral Galaxies, IV, Systematic

Dynamical Properties, Sa → Sc. *The Astrophysical Journal, 225*, L107–L111. https://articles.adsabs.harvard.edu/pdf/1978apj. . .225l.107r

Stewart, G. (2025, June 5). *New Quarter Honors Vera Rubin, Astronomer Who Revealed the Universe's Hidden Mass.* Smithsonian American Women's History Museum. https://womenshistory.si.edu/blog/new-quarter-honors-vera-rubin-astronomer-who-revealed-universes-hidden-mass

ThemeGrill. (n.d.). *Vera Rubin: A life – British Astronomical Association.* Retrieved June 24, 2026, from https://britastro.org/journal_book_review/vera-rubin-a-life

Timeline - Women in exploration. (n.d.). *Vera Florence Cooper Rubin.* Retrieved June 24, 2026, from https://womeninexploration.org/timeline/vera-florence-cooper-rubin/

*Vera Rubin | Biography | Research Starters | EBSCO Research.* (n.d.). EBSCO. Retrieved June 24, 2026, from https://www.ebsco.com/research-starters/biography/vera-rubin

*Vera Rubin.* (n.d.). National Women's History Museum. Retrieved June 14, 2026, from https://www.womenshistory.org/education-resources/biographies/vera-rubin

*Vera Rubin.* (n.d.-b). Biographs. Retrieved June 24, 2026, from https://biographs.org/vera-rubin

*Vera Rubin: A Life, with Author Dr. Jacqueline Mitton - AIP.ORG.* (2025, October 28). AIP. https://www.aip.org/library/vera-rubin-a-life-with-author-dr-jacqueline-mitton

*Vera Rubin: Opening doors to dark matter and women in STEM.* (2024, October 9). Carnegie Science. https://carnegiescience.edu/vera-rubin-opening-doors-dark-matter-and-women-stem

*Who was Vera Rubin? — Harvard University Press.* (n.d.). Harvard University Press. Retrieved June 14, 2026, from https://www.hup.harvard.edu/features/who-was-vera-rubin

*Who was Vera Rubin?* (2016, December 15). Carnegie Science. https://carnegiescience.edu/news/remembering-vera

Wikipedia contributors. (2026b, May 15). *Vera Rubin*. Wikipedia. https://en.wikipedia.org/wiki/Vera_Rubin

Yeager, A. (2008). *Cosmos Incognito: Vera Rubin shines light on dark matter.*

# ILLUSTRATION CREDITS

Every effort has been made to identify and appropriately credit all photographs, illustrations, and other visual materials used in this publication. Original illustrations are credited to their creators, and third-party images are reproduced or adapted in accordance with their applicable licenses or public domain status.

### Marie Curie:

Original portrait illustration by Osmo, adapted from a historical photograph by an unknown photographer (c. 1898), in the public domain.

### Marie and Pierre Curie in their laboratory:

Historical photograph by unknown photographer (c. 1904) via Wikimedia Commons. Public domain.

### Jane Goodall:

Original portrait illustration by Osmo, adapted from a photograph by Al Hartmann for *The Salt Lake Tribune* (c. 1986), used under the Creative Commons Attribution 4.0 International (CC BY 4.0) license.

### Two young chimpanzees in Gombe National Park, Tanzania:

Photograph by Cethuyghe. Licensed under Creative Commons Attribution-ShareAlike 4.0 International (CC BY-SA 4.0). Adapted for publication in black and white.

## Ada Lovelace:

Original portrait illustration by Osmo, adapted from a historical portrait by Antoine Claudet (c. 1843), in the public domain.

## A page from Ada Lovelace's 1843 Notes on the Analytical Engine:

Reproduced from *Scientific Memoirs*, Vol. III (1843), public domain, via the Internet Archive.

## Rosalind Franklin:

Original portrait illustration by Osmo, based on a photograph by Vittorio Luzzati (c. unknown) for Cold Spring Harbor Laboratory, used under the Creative Commons Attribution-ShareAlike 4.0 International (CC BY-SA 4.0) license.

## DNA diffraction illustration:

Original illustration by the author, inspired by Rosalind Franklin's pioneering X-ray diffraction research on DNA.

## Vera Rubin:

Original portrait illustration by Osmo, adapted from a photograph by Emilio Segrè (c. unknown), courtesy of the AIP Emilio Segrè Visual Archives, American Institute of Physics, used under the AIP Attribution License.

## Galaxy rotation illustration:

Original illustration by the author, inspired by the pioneering galaxy rotation studies of Vera Rubin and Kent Ford.

# ABOUT THE AUTHOR

R. D. Bennett is the author of the *Women Who Did* series, which celebrates the remarkable women whose ideas, discoveries, and determination changed the world. Through careful research and engaging storytelling, she brings history to life for readers curious about the people behind some of humanity's greatest achievements.

With a background in healthcare and leadership, she is passionate about making science and history accessible to everyone and inspiring the next generation of innovators, explorers, and problem-solvers.

She lives in Virginia with her family and two dogs, where she continues to write books that educate, inspire, and encourage lifelong curiosity.

# THE WOMEN WHO DID SERIES

History is filled with extraordinary women whose ideas changed the world. Through engaging storytelling and carefully researched biographies, the *Women Who Did* series introduces readers to the pioneers, explorers, inventors, scientists, and changemakers who shaped our past and continue to inspire our future.

*The journey has just begun.*

Coming Soon

# WOMEN WHO DID EXPLORATION

Discover the remarkable women who crossed oceans, climbed mountains, mapped unknown lands, and expanded our understanding of the world.

www.ingramcontent.com/pod-product-compliance
Lightning Source LLC
LaVergne TN
LVHW050909210926
848482LV00012B/380
* 9 7 9 8 9 9 2 6 7 9 7 9 3 *